Financial Justification of Nondestructive Testing

Cost of Quality in Manufacturing

Emmanuel P. Papadakis

CRC Press
Taylor & Francis Group
Boca Raton London New York

CRC Press is an imprint of the
Taylor & Francis Group, an **informa** business

Published in 2007 by
CRC Press
Taylor & Francis Group
6000 Broken Sound Parkway NW, Suite 300
Boca Raton, FL 33487-2742

International Standard Book Number-13: 978-0-8493-9719-6 (hbk)
Library of Congress Card Number 2006008691

Library of Congress Cataloging-in-Publication Data

Papadakis, Emmanuel P.
Financial justification of nondestructive testing : cost of quality in manufacturing / Emmanuel P. Papadakis.
p. cm.
Includes bibliographical references and index.
ISBN 0-8493-9719-7
1. Nondestructive testing--Cost effectiveness. I. Title.

TA417.2.P37 2006
658.5'68--dc22 2006008691

Preface

The principal impetus for the writing of this book is the author's realization that financial calculations provide the key to the implementation of nondestructive testing (NDT) for improved quality in industrial output. Scientists and engineers in industry generally have not learned much finance in their formal educations and are at a loss to be able to prove financially that their proposals for new methods and equipment are justifiable. These scientists and engineers are experts in the technical methods needed to accomplish projects. This is equally true of NDT specialists and engineers in other specialties. They generally know how to improve quality but do not know how to prove that their improvements will make money for their employer. The engineers are generally at a loss when it becomes necessary to demonstrate, to their own management and to higher management such as controllers and treasurers, that their methods are justified quantitatively on the basis of making money for the company or saving money for the government office. This book is intended to show the scientists and engineers how to justify their NDT project on the basis of finance.

A derivation in an early version of Dr. W. E. Deming's main book (Deming, 1981, 1982) led the author to study the question of quantitative finance as a way to choose to test or not to test manufactured product on the production line. This study branched out into the case of the need to analyze investments in inspection equipment that was to be used for more than 1 year on a project. When several years were involved, the question of profit and loss over time was raised. Deming's idea of staying in business and improving competitive position led to another formulation of the costs of testing and the cost due to nonconforming material in the big picture of quality. This book puts it all together by teaching three methods of making financial calculations to prove or disprove the need for the long-term use of 100% inspection. The author's introduction to finance came through a master's in management (1979) from the University of Michigan under the sponsorship of the Ford Motor Company. He had two semesters of economics and three semesters of accounting, among other courses, and also studied TQM under W.E. Deming and W.W. Scherkenbach at Ford.

Introduction

This book introduces the concept that 100% inspection using high-tech methods can save money for a manufacturing organization even though the inspection itself adds a modicum of cost to the manufacturing. Three methods of calculation will be taught to justify the use of high-tech 100% inspection. The saving of money arises through the elimination of nonconforming material from the output. The operative principle is that the detrimental cost to the organization of one nonconforming part's escaping into the field (being sold to a customer) can be enormous compared with the cost of the part itself and gigantic compared with the cost to test it. In some cases the detrimental cost (also called disvalue in this book and value-added-detractor [VADOR] in telephone system parlance) can be so large that just a few nonconforming parts can change the picture from profit to loss for a manufacturing process. Financial calculations are the court of last resort in all those cases in which no overriding simplistic arbiter of testing is present.

Let us first investigate what is meant by a "simplistic arbiter." A simplistic arbiter is any statement that can be written as "you must" or "the organization shall" do testing. Such statements may arise from laws and their interpreters such as the National Transportation Safety Board (NTSB), the Federal Aviation Administration (FAA), and the like including military organizations that must keep equipment operational. Statutory and regulatory demands must be met. Such statements also may arise from firm commitments to organizations such as the International Standardization Organization (ISO) with its all-encompassing set of ISO standards. These are simplistic in the sense that if an organization chooses to adhere to them or is forced to obey them, then the decision as to testing or not testing is made for the organization and is no longer subject to discussion or justification.

Other cases of arbitrary imposition of testing rules arise from court cases. One famous case showing the limitations of financial calculations for making engineering choices is the Pinto Fire, in which the automobile manufacturer chose to save money by omitting a safety shield in the vicinity of the gas tank. The financial calculation used in those days balanced the loss expected from lawsuits for wrongful deaths against the cost of installing the safety devices on all the cars of that type made. The corporate estimate of the cost of a life was about $500,000. However, when a young lady was burned to death in a car struck from behind, the judge awarded $125,000,000. The judge also ordered that the cost of a life should never be included in the cost-benefit calculation, but rather that the manufacturer should do anything a reasonable person would do within the state of the industry to eliminate the danger. This became a benchmark for the NTSB in automotive cases. ("State-of-the-industry" is what you can buy from a vendor; it may not be as good as "state-of-the-art," which has just been reported at a scientific society meeting.) In the Pinto fire case, this meant installing the safety shield at a cost of about $2 on each car. While this case could not have been solved by installing nondestructive testing (NDT), the concept turns out to be very relevant for deciding about NDT in production. Concerning manufacturing flaws in safety items, a senior lawyer at the Office of General Council of the automobile company explained the situation as follows: If a flaw in a safety-related part is discovered in the field (i.e., after a car has been shipped from the factory), then it is required of the manufacturer to do whatever a reasonable person would be expected to do to ensure that this flawed part is unique in the universe. Now, "what a reasonable person would do" and "unique in the universe" are terms exactly defined and understood in law. The law does not say that you have to do NDT; neither does it say that you have to do statistical process control, or possibly something else. The firm has the choice. The choice depends on probability of detection, Type I versus Type II errors, and costs. While it may be impossible for either state-of-the-industry NDT or statistical process control (SPC) to ensure that no defectives will ever be produced in the future, it is incumbent upon the industry to choose the best method and do whatever a reasonable person would do to rectify the situation now and in the future. This might include NDT research. Incidentally, it should be pointed out that implementation of NDT is complicated and hampered by dogmatic positions taken by statisticians. One tenet of statisticians is that reliance upon inspection should be eliminated. This unscientific approach will be discussed later in Chapter 4.

Inspection by means of NDT is a process that has a definite place in the big picture of quality. Finance is a major key to the implementation of NDT in production. NDT personnel must be able to justify the use of NDT by means of the financial calculations to be given in this book. Only then will they be able to convince their controllers and financial officers to expend the resources to set up and run the necessary NDT inspections.

The idea of "inspection" has frequently been understood in terms of one of W. E. Deming's Fourteen Points, which states, "Cease dependence upon mass inspection." By some quality professionals this is translated illogically into an action item that advocates the elimination of all inspection including NDT to get rid of the alleged addictive qualities of inspection, and then the substitution of such a high central capability in the manufacturing process as to make inspection unnecessary. Reaching the high process capability is to be accomplished, according to the statisticians, by "continuous improvement." The statisticians believe that "continuous improvement" will eliminate the need for inspection. On the basis of this credo, the statisticians deprecate and eliminate NDT *a priori*. This denigration of NDT is illogical for two reasons:

1. The addiction to inspection grips a company only if the engineering management
   (a) Fails to use the results of inspection in a timely fashion.
   (b) Uses the inspection as a crutch to eliminate nonconforming material without fixing the process.

   This sort of management behavior is lazy as well as improvident, and should be eliminated anyway.
2. High capability of manufacturing processes may not be adequate to eliminate the need for inspection.

With increasing capability, the condition of "small number statistics" is approached where even the small proportion of nonconforming output might not be detected by statistics and could still have catastrophic consequences. Moreover, some kinds of nonconformities can be found only by means of NDT technologies. In addition, engineering will frequently tighten specifications or introduce more difficult designs just because they notice that the manufacturing capability has become higher, automatically making the capability lower again. The understanding of these concepts as taught in this book is necessary for NDT professionals in manufacturing who must address inspection issues. The NDT personnel should learn the financial calculations to be given in this book. Other quality professionals would benefit as well.

Management philosophies and mindsets that led to the improper dependence upon mass inspection are analyzed. This is a necessary background to understand where the inspection people and the statisticians are coming from in their present-day confrontation concerning NDT. The Taylor management philosophy of kicking all decision making "upstairs" and treating all workers as just hands (no brains) is shown to be the principal culprit. Present-day methodologies such as total quality management (TQM) and standards such as ISO-9000 are shown in their proper relationship to quality. How inspection by NDT fits into them is explained clause by clause. The role of NDT as a means of inspection is shown. The professionals in NDT and the management of the quality function in an organization all need a firm understanding of this melding. In this book, NDT will be emphasized when 100% inspection and/or automated inspection is referred to, although there are other valid methods of inspection such as laser gauging that can handle some situations and be automated and applied to 100% of production. Occasionally, the NDT must be performed by technicians using equipment rather than by automation. Financial calculations to be taught involve both investments and variable costs.

When one analyzes the corporate addiction to inspection and the proposed over-compensation by means of manufacturing process capability, the net result is that 100% inspection by NDT may be necessary indefinitely or for protracted periods of time until it can be proven to be unnecessary. The word "proven" is operative here. One must understand this concept of proof to function effectively at the interface of inspection with the rest of the quality system. There are ways and means to prove financially that inspection of 100% of production (particularly by NDT) **should** be performed, or that it **should not** be performed. The assumption here is that the presence of nonconformities has only financial implications. (See below for comments about health and safety.) This book presents **three major methods** for financial calculations to **prove or disprove** the need for 100% inspection. Plentiful examples are drawn from case studies of NDT used in inspection in manufacturing industries.

There are situations in which health is at risk that require 100% inspection forever no matter what the capability of the process. These situations are explained. Also, there are situations in which 100% inspection should be carried on for information-gathering until a process is proved capable and stable. These situations are recapitulated.

It is emphasized that processes must be brought under control and kept under control before the financial calculations on the continuing need for 100% inspection by NDT can be performed in a valid manner for the long term. To do this, SPC is advocated. A functional review of SPC is presented with deference to the many good books on the subject.

Then the three financial methods for calculating the need for 100% inspection are presented. NDT personnel will find them instructive and useful, and to the NDT professional they will become second nature. The financial methods are (1) the Deming inspection criterion, which is particularly useful for cases involving small capital investments and larger variable costs; (2) the time-adjusted rate-of-return or, almost equivalently, the internal rate of return calculation, which is useful for cases involving large capital investments used over several years; and (3) the productivity, profitability, and revenue method pioneered by this author, in which productivity is written in terms of dollars-in vs. dollars-out for any process. The productivity method can be considered as nano-economics for all processes within a firm. The sources of adverse costs to a firm from nonconformities are addressed. Also, the sources of testing costs are listed.

The three financial methods can prove that 100% inspection by NDT methods is actually **profitable** to a firm under certain circumstances despite high capability and process-under-control conditions. Examples are drawn from successful uses of NDT methods as the means of inspection. This exposition of the methods and their calculations makes it possible for the NDT engineer or Level III technician, the statistician, the quality engineer, the company controller, the treasurer, the manufacturing manager, the CEO, and anyone else with responsibility or authority to compute the advisability of using 100% inspection. No other book performs this necessary task.

Many examples from the real world of engineering and manufacturing are presented to illustrate the financial methods. Both positive and negative decisions about NDT used for 100% inspection are shown. Cases are given in which 100% inspection remained necessary even after periods of diligent "continuous improvement."

Cases of inspection that increased corporate profits by millions of dollars a year while costing only a few thousand dollars are presented. Some cases where newly invented NDT inspection methods averted catastrophes in major corporations are set forth. Improper management decisions **not** to install inspection are addressed. Some but not all of these examples are in technical papers scattered through the literature. Only in this book are they presented as a succinct unit.

The conclusion is that 100% inspection by NDT and other valid techniques has a rightful place in the set of methods used by quality professionals. The decision to use or not to use 100% inspection can be made rationally on a financial basis within the working context of SPC and high capability. The methods for making the decision are enunciated. Managers, quality professionals, NDT specialists, and inspection technologists need this book. Students entering the field will find it invaluable.

Notes on How To Use This Book

For a person who wishes to address the question of the financial justification of the application of inspection by nondestructive testing (NDT) to 100% of production in manufacturing without delay, he should read the theory in Chapter 7 and the applications in Chapter 9. Then using Chapter 6, he will be able to recognize the methods of putting cost data into the financial equations and solving them for the YES/NO answer to the question of testing. The person will have to study his own company to find the actual dollar values and production data to insert. The part of Chapter 6 on the need for statistical process control (SPC) to be used in the production process must be read because SPC is a prerequisite to ensure that the process is under control during the times the data are taken for the financial equations. The person totally familiar with SPC will find this synthesis satisfying.

The person not familiar with SPC will benefit from the longer explanation of it in Chapter 3. This chapter is basically only a beginning of the study of SPC, which should be pursued using the references cited and other courses offered in various institutions. Chapter 3 is really an introduction to the subject of SPC for technical personnel not familiar with the work of quality professionals.

Many people will be familiar with SPC but not conversant with NDT. A number of examples of different types of NDT are introduced in Chapter 8 as high-tech inspection methods. It is hoped that the brief descriptions of the methods will give the reader the insight to see that there are many methods available and others to be invented. One does not need to be an engineer, scientist, or mathematician to use these methods. Basically, one calls a salesman for a reliable company making the equipment or an NDT consulting firm and plans an approach to fit the problem.

As a background for the need for systematic efforts to improve quality, Chapter 2 traces the development of industry from its beginning through the implementation of mass production. One of the final formulations, Scientific Management, also known as Taylorism, is addressed at length because the following wave of manufacturing philosophy, total quality management (TQM), has tended to lay the blame for poor quality at the feet of Frederick Winslow Taylor who set forth the principles of Scientific Management. The situation seems to be that the results were not as salutary as the intent of Taylorism. Taylor stated and implemented a philosophy specifying how people should be organized and how people should be treated to maximize their output, productivity, and efficiency in particular.

TQM is introduced in Chapter 4 stressing in particular the ideas of W. E. Deming. TQM is a philosophy stating what people should do and how people should be treated to have, as a result, good quality in the output of their firm. It advocates the position that firms create poor quality by failing to correctly manage their employees as well as various aspects of their business. TQM generally incorporates SPC as a prerequisite. Certain TQM misunderstandings about NDT are reviewed because it is necessary for the practitioner of quality improvement to understand the interaction of TQM and inspection technology.

The most recent attempt to systematize the production of high-quality goods is the ISO-9000 quality management standard. Its development and implications are outlined in Chapter 5. The progression of the standard is toward the emphasis of TQM, but there are opportunities for company management to implement 100% NDT of production correctly even in this context.

The student approaching this subject for the first time will benefit by starting at the beginning and going straight through.

The quality professional and the high-tech practitioner in the field of quality should absorb this book in its entirety. Manufacturing would be the better for the effort.

Author

Emmanuel P. Papadakis, Ph.D., is president and principal in Quality Systems Concepts, Inc. (QSC), a firm in quality and nondestructive testing (NDT) consulting. He has been a provisional quality auditor under the Registered Accreditation Board (RAB) system. He received his Ph.D. in physics (1962) from the Massachusetts Institute of Technology (MIT) and his master's in management (1979) from the University of Michigan. Before QSC, he was associate director of the Center for Nondestructive Evaluation at Iowa State University. Prior to that, he managed research and development (R&D) in NDT and product inspection at the Ford Motor Company, leading a group that expanded its work from R&D in NDT to include product quality research with statistical systems and financial analyses of NDT culminating in quality concepts for new vehicles. While at Ford, he served on the Statistical Methods Council that W. E. Deming set up to implement his philosophy at the Ford Motor Company. Dr. Papadakis previously served as department head of physical acoustics at Panametrics, Inc., where he managed government R&D, private consulting, product development, and transducer design. Before that, he was a member of the technical staff at Bell Telephone Laboratories, where he worked on sonic and ultrasonic devices and associated fundamental studies on materials, wave propagation, measurement methods, and NDT. He got his start in ultrasonics and NDT at the Watertown Arsenal during graduate work at MIT, where his thesis was in physical acoustics and solid state physics, dealing predominantly with ultrasonics.

Acknowledgments

Many people have provided invaluable help with this volume.

First, I want to thank my wife Stella for her patience while I was spending so much time on the process of writing, and, even more, on the process of thinking, which takes time and concentration away from more light-hearted endeavors. Stella has been helping me ever since typing my thesis in 1962.

My brother Myron helped over a period of several years with insights into product liability law. He wrote one section in this book detailing the need for continuity in engineering knowledge to illustrate the possibilities of calamities when former knowledge is forgotten.

My father, quoted posthumously, provided some oral history by way of dinner-table conversations that proved very relevant to describing the milieu of factory work early in the twentieth century.

Arthur J. Cox provided some famous as well as some obscure texts and letters elucidating the development of manufacturing in America up through mass production. His book on the Ferracute Manufacturing Company will be of interest to scholars studying individual companies.

Charles E. Feltner at the Ford Motor Company supported my professional involvement as well as my industrial work as my department manager for several years. He provided incentives to learn more about nondestructive testing (NDT) beyond ultrasonics and more about quality beyond NDT. Feltner was instrumental in assigning me to the Deming classes at Ford and to membership in William W. Scherkenbach's Statistical Methods Council, which Deming set up there to oversee total quality management (TQM) and statistical implementation. For my part, I was eager to follow this direction. Many of the financial examples of NDT justification cited in this book come from my work on warranty questions and other quality concerns I encountered while supervising a section on NDT and quality in Feltner's department.

Craig H. Stephan of my section in Feltner's department helped by supplying information and reprints on case depth by eddy current correlations. Gilbert B. Chapman II, also of my section, provided updated information on infrared applications and evanescent sonic waves. Stan Mocarski of another development group provided necessary data during concurrent engineering sessions.

David Fanning, editor of *Materials Evaluation* at the American Society for Nondestructive Testing (ASNT), searched numerous references, names, and phone numbers.

Conversations with William W. Scherkenbach, G. F. Bolling, Rod Stanley, and Bruce Hoadley proved enlightening and helpful.

Fletcher Bray and Tom Howell of the Garrett Engine Division of the Allied Signal Aerospace Company supplied data on quality of jet engine discs while I worked with their company as a member of the Center for Nondestructive Testing at Iowa State University. The disc data proved invaluable in the financial analyses in this book.

Work with H. Pierre Salle of KEMA Registered Quality, Inc., broadened my knowledge of ISO-9000.

I am grateful to Thrygve R. Meeker who, earlier in my career, mentored me in professional pursuits in the Institute of Electrical and Electronics Engineers (IEEE) group on ultrasonic engineering and in the Acoustical Society of America.

My son, Nicholas E. Papadakis, created the digital files for the drawings and photographs in the book.

Contents

1

The Big Picture of Quality

1.1 What Quality Means to People

It should be stated at the outset that the principal subject of this book is financial calculations to prove or disprove the need for 100% inspection of manufactured goods by high-tech methods, particularly nondestructive testing (NDT). The intent is to fit this calculational methodology into the entire context of quality so that management as well as NDT professionals will feel comfortable with it. To do this, it is necessary to provide a background to the overall "big picture" of quality. This first chapter provides some of this background, and Chapter 2 continues the exposition from a historical perspective showing how inspection came into the quality picture.

Precisely because quality is qualitative, quality is very elusive to describe. However, quality managers are among the first to attempt to express quality in quantitative terms despite its qualitative nature. Personnel dealing with quality will quantify process capability, control limits, average outgoing quality, specification limits, and a host of other vocabulary with quantitative meanings to try to express quality in the arcane and ever-changing diction of the day. These concepts are useful and even necessary to keep industry running and turning out material that customers will buy, but they beg the main question. Can this question be stated?

The principal question is, "What will people be willing to buy?" This question leads into the concepts of fitness for use, value, and most important, the perception of quality. Quality is precisely that, namely, a perception. Individuals have a perception of quality, which is an expression of what they think of as good. If you ask a person a question such as, "What makes a good pancake syrup?" you may get any number of answers including some as specific as "Vermont maple syrup" or even "Vermont Grade A Light Amber maple syrup." Back at the food processing plant these quality specifications are quantified, of course, by density, viscosity, colorimetry, boiling point, source of the sap, or other raw materials, such as winter snow and spring thaw temperatures, quality parameters of all the raw materials, a manufacturing process, and so on. If the product is a mixture, there is a formula or recipe, too, and a process. Much quantitative work goes into quality.

TABLE 1.1

Personal Perceptions of Quality

PERSON	OPINION
1. Businessman	"Features"
2. Club woman	"Accessories"
3. Professional	"Good workmanship"
4. Hot rodder	"Performance"
5. Farmer	"Durable"
6. College boy	"Well put together"
7. College girl	"Beautiful"
8. Military officer	"Reliable"
9. Engineer	"Meets specifications"

Back to people's perception of quality. Some people believe wholeheartedly that one kind of syrup is "better" than another. I was amazed to find one fine hotel in Boston serving the "Vermont Grade A Light Amber maple syrup" on halves of cantaloupe, filling the hemispherical hole. A large chain of roadside restaurants in the Southeast and the Midwest serves "Vermont Grade A Medium Amber maple syrup" on pancakes and waffles. You get as much as you want in little bottles the same size as used for liquor on airliners. The restaurant chain claims that it is the largest consumer of Vermont maple syrup in the world. But on the other hand I know individuals who say that real maple syrup is unacceptable; they want cane sugar-based pancake syrup.

Perception of quality is just as varied in any other industry. Suppose a group of diverse individuals was to be asked what constitutes a good automobile. Table 1.1 shows a range of answers. Only the engineer at the end of the table says anything mathematical or strictly quantitative. The array of answers means that the marketing function of a company must find out what should be produced before it turns the product idea over to the myopic designers and engineers.

An elderly gentleman was once asked what kinds of cars he had purchased throughout his lifetime. His answer was, "Brand X and Brand Y, alternately, about every 6 years." When asked why he did not ever purchase Brand Z, he answered, "Because it isn't good enough." He didn't even consider Brand W. Now, throughout his lifetime, Brand X, Brand Z, and Brand W were competing head-to-head, and Brand Y had a much smaller share of the market although it had a reputation for craftsmanship. Without more data, we can say only that the gentleman had a perception of quality, value, and fitness for use, which gave him definite opinions about automobiles. There is one more lonesome data point about the quality of Brand Y. The gentleman stored a 1922 touring car by Brand Y in a barn on his farm when he bought a new car later that decade (Eastman, 1947). He did no service on it while in storage. In 1946 a soldier returning from World War II bought the car, added fluids, and drove it away. This author knew the gentleman, the soldier, and the car. The gentleman was a highly accomplished engineer and entrepreneur in steel erection.

The gentleman above provides a snapshot of the perception of quality from one perspective. The Big Picture of Quality requires four things: (1) the supplier determines the desires of the customers across the entire scope of the set of answers, such as those given in Table 1.1, taking into consideration value and fitness for use; (2) the supplier designs and builds goods that actually fit the wants of the customers; (3) the supplier controls his manufacturing mechanisms to keep producing the desired output; and (4) the supplier is capable of proving that the control is both ongoing and applied to all phases of the business. These requirements point to the fact that quality must be managed. How is this elusive requirement to be met?

1.2 Trying To Manage Quality

The attempt to manage quality has gone through several stages and has produced many solutions depending upon the assumptions made concerning production.

Before mass production, the master craftsman controlled his journeymen and apprentices by on-the-job training and visual inspection, both end-of-line and verification-in-process as we would say today.

Under early mass production with interchangeable parts, vast systems of jigs and fixtures as well as gauges were used (and still are) to ensure that things fit. Concurrently, the quality of supplies from suppliers who were essentially "black boxes" needed checking. The buyer had no control over the seller except the threat to cease making purchases. There was a need to know whether material bought over-the-counter was good enough to use in the purchaser's goods. Statistical methodologies were developed in great detail to determine the percentage of nonconforming material in batches and the probability that batches contained no nonconformities (Shewhart, 1931; Western Electric Co., 1956). These methodologies were applied widely in incoming inspection, a necessary management function of that period. The situation was that one did not know whether the supplier's process was under control and the further assumption was that there was no way of knowing. Another situation was that one's own processes might go out of control and not be detected. It was a truism that the final detection might not happen for protracted periods of time, allowing mountains of nonconforming production. The assumption was that the out-of-control conditions could not be detected in a timely fashion. Thus, internally to a company, inspection was mandated and played a major role in the quality of outgoing product. Finally, the concept of a process going out of control was recognized. It was addressed in several ways.

Initially, statisticians developed methods to determine probabilistically whether processes were actually under control (Shewhart, 1931). The result, statistical process control (SPC), if applied, was useful to a company for its own prudent management of resources but was of little use in commerce. This failure was due to two situations: (1) company secrecy about new methods, and (2) lack of control of a purchaser over his supplier in free commerce. Each of them was an independent entity making an arm's-length transaction. The purchaser took bids on price, and the cheapest supplier won. (It was just like you, an individual, buying a house from another person. The two make a deal when one is ready, willing, and able to sell and the other is ready, willing, and able to buy.) A thoroughgoing application of SPC might have resulted in the lowest prices in the industrial deal case, but this was not recognized until the 1980s. Meanwhile, SPC did find uses in some companies that were strongly vertically integrated. AT&T comes to mind (Western Electric Co., 1956). Knowing that the supplies from Western Electric (the wholly owned sole supplier) for use by Long Lines (the long distance division of AT&T) were good material from controlled processes was valuable in the context of the vertical integration of the Bell Telephone System.

More recently, it has become the practice in some very large companies to put the responsibility of good production onto the shoulders of the suppliers by teaching them SPC, insisting upon its use, and requiring documentation daily (Automotive Industry Action Group, 1995). This is a monumental task but its advocates claim success. Their "police," the old Supplier Quality Assurance branch, now can be constructive as Supplier Quality Assistance and can live under a different cooperative corporate culture.

Each group of the above has tended to advocate its approach as the only sound one.

The major development in quality management in the past decade has been the International Standardization Organization (ISO)-9000 quality management standard. This different approach is explained in the next section.

1.3 ISO-9000 as the Management Standard for Quality (Revised 2000)

ISO-9000 has been pioneered by the European Economic Community as both a method to enforce its own unity and as a method to require that the world meet its standards in order to continue to trade with it. One of the community's principal driving forces has been concern over the quality of medical supplies and equipment. It seems as worried over adhesive plaster as the United States became over thalidomide. However, the whole world has jumped on the ISO-9000 bandwagon so that the fear of monopoly and boycott has passed without incident. The world is now into the second round of ISO-9000, namely ISO-9000–2000. The difference between the 1990 version and the 2000 version will be shown by first explaining the 1990 version and then showing the changes made in the 2000 version.

Companies and institutions throughout the world have become quality registrars to enroll other organizations into the fraternity, so with effort, any company can achieve registration and hence access to markets. In the process of becoming registered, the companies are supposed to get a better handle upon quality and possibly even improve. Their registrar authority must audit them periodically to ensure compliance. The year 2000 version of ISO-9000 specifies the need for improvement. These procedures will be explained in their own chapter.

So what is a quality management standard in this sense? The ISO-9000 standard specifies generic activities for all functions of any organization to keep these functions operating successfully day-in and day-out. The basic assumption of the 1990 version of the quality management standard is this, colloquially: "If the organization is functioning today with adequate quality to satisfy its customers, then following the ISO-9000 quality management standard will assure that the organization will continue to produce adequate quality." The quality management standard is principally concerned with proof that the organization has performed in the way it has promised to perform. It does this by specifying audits of ongoing performance. As mentioned, ISO-9000–2000 mandates improvements, as well. Statisticians are gaining a greater degree of control as time goes on.

The promises and the proof are set forth in a hierarchical set of requirements in five tiers. All of these must be documented.

1.3.1 Five Tiers of Quality Management per ISO-9000

1. First Tier: The company should have a vision statement that calls out quality as a goal. This is a quality policy. It is a document the chief executive officer (CEO) and every other important officer signs.
2. Second Tier: The second tier is a quality manual that addresses all the items and operations that can affect quality in the company operations. All the topics in the ISO-9000 quality management standard must be addressed.
3. Third Tier: The third tier is a set of standard operating procedures for every aspect of the company business and, in particular, for every process that takes place in the company.
4. Fourth Tier: The fourth tier is a set of detailed work instructions for operators to follow in running every process that goes on in the company. Nothing is left to chance, education, or intelligence.
5. Fifth Tier: The fifth tier is a compendium of quality records in which every operator writes down and acknowledges that the work instructions were carried out daily. Records of other variables such as temperature, humidity, brown-out voltages, and every conceivable perturbation would also be recorded. These records are to be available for internal and external audits to show that the instructions were carried out continuously.

TABLE 1.2

ISO-9000 Quality Management Standard: 1990 Issue

Full Version 9001 (For Organizations Including Design Functions) Table of Contents of Part 4
1. Management Responsibility
2. Quality System (Quality Manual)
3. Contract Review
4. Design Control
5. Document and Data Control
6. Purchasing
7. Control of Customer-Supplied Product
8. Product Identification and Traceability
9. Process Control
10. Inspection and Testing
11. Control of Inspection, Measuring, and Test Equipment
12. Inspection and Test Status
13. Control of Nonconforming Product
14. Corrective and Preventive Action
15. Handling, Storage, Packaging, Preservation, and Delivery
16. Control of Quality Records
17. Internal Quality Audits
18. Training
19. Servicing
20. Statistical Techniques

The table of contents of Part 4 of the 1990 version of the standard is given in Table 1.2 and lists all the sections within the second tier for a company using the 1990 version. (Parts 1 through 3 of the standard are completely administrative and not technical.) These sections in Part 4 generally cut across departmental lines. The third tier calls for a complete set of written procedures for all processes, both engineering- and management-oriented. At the fourth tier, every procedure must have a set of unambiguous Work Instructions for the operators to follow in the factory, laboratory, office, shipping dock, etc. The bottom tier of the pyramid is quality records, which is a system of additional documents that are filled out, signed off on, and stored to show that all the work instructions were followed. For all the above documents, the latest versions must be available at the workstations and the old versions must be discarded to eliminate ambiguity. (The quality manager may keep archival copies, of course.) This entire set of documents and documentation constitutes what the organization has promised to do and the proof that it has performed as promised. The method of enforcement is through periodic audits of these documents and the workplace by the quality registrars.

It is important to note the following proviso or limitation. The standard does not specify the content of any organization's promise to itself or its customers. The organization is not told how to run its business. It is simply told to keep running the same way as always and prove it. For instance, the standard does not specify SPC for keeping processes under control. The standard simply asks for proof that statistics is being used if the organizational plan calls for statistics. The standard does not call out the use of any particular type of measuring device. The standard does, however, ask for assurances that the organization use instruments as the organization's plan specifies and that the organization keep the instruments calibrated. The organization must be able to prove that the calibration is done traceably as frequently as the organization's plan calls for, and so on for all the quality-related items one can rationally think up. The standard is so thorough that it even talks about preserving input/output shipments from corrosion.

ISO-9000 alludes to the use of quality methodologies that are currently in use by quality professionals. There is very little in the way of prescription or proscription. Parts of Chapter 5 in this book analyze how some of the clauses in a few of the sections in the standard impact the question of inspection technology, and, in particular, NDT. Understanding these clauses will be critical for the quality professional and the NDT specialist. The Year 2000 version has some new wording to attempt to introduce proactive total quality management and particularly continuous improvement. How this works out will have to be seen by experience. Certain industry-specific derivative quality standards include even more emphasis on continuous improvement, SPC, and specific methodologies such as failure mode and effects analysis. These industry-specific standards are beyond the scope of this book.

2

How We Got to Where We Are

2.1 Early Philosophy of Manufacturing

Early manufacturing was carried out by journeymen and apprentices under the supervision and tutelage of master craftsmen. The masters negotiated, designed, and directed while the journeymen did most of the crafts work and the apprentices were labor, gofers, and power sources. For instance, an apprentice in a woodworking shop would have to turn the giant wheel over which a leather belt sped along turning the lathe. The journeyman held the tool to turn the chair leg, for instance, on the lathe. The master would judge if the two front legs for the chair turned out similar enough. If water power were available, the job of the apprentice might be easier. Apprentices were usually indentured servants for a period of 6 to 10 years. They were supposed to look over the shoulder of the journeyman to learn the trade. Some teaching (on-the-job-training) went on, as the master wanted the apprentice to be promoted to journeyman at the end of his indenture. An industrious father would want his son to be indentured to a good master who would bring the boy up into the business. The boy's hard work was considered training, not child labor.

A good master could become quite wealthy and even famous for his wares if he were in the right business at the right location at the right time. The names Chippendale, Hepplewhite, Sheraton, Pfyfe, Goddard, Townsend, Hitchcock, and Terry come to mind. These and others made superlative products, which are now heirloom and museum quality. Books were written by them and about them, and continue to be written and reprinted today. See, for instance, Chippendale (1966) and Sack (1950). Hitchcock and Terry slide over into the modern manufacturing era as well as representing ancient craftsmanship. Hitchcock as a traveling salesman from Connecticut sold chairs as far west as the little town of Chicago in the early 1800s.

Certain technologies involving craftsmanship were brought to high levels of skill by requiring each master to adhere to the standards and regulations of his guild. The craft passed from the master "professor" to the journeyman "graduate student" to the apprentice "student." To be able to do this, the master had to work up to a point where he owned a small business and employed a staff of journeymen and apprentices. This and all smaller establishments were termed "cottage industry" in modern parlance.

A guild was essentially a slave master to the masters. He could not leave to set up shop elsewhere. The guild hoped to keep a monopoly on some technology by restraint of trade. A well-known example is fine glass blowing. When some German experts escaped and came to America, there was an explosion in glass technology.

At some point, specialization entered into the making of things within a single shop. An interesting fictional but believable account of the invention of specialization in hand-manufacturing is in the novel *Les Misérables* (Hugo, 1862/1976). The hero, after escaping from a prison trireme, had obtained a job in a pottery factory. Before his arrival, each potter did all the operations such as mixing clay, turning vessels on the potter's wheel, and painting the floral decorations before firing. The hero noted that one man was excellent at turning and another was excellent at painting. He arranged for these men to become specialists. Quality improved, production increased, and profits went up. Unfortunately, the hero was apprehended by the cruel French detective. *Sic transit gloria.*

At this point the making of things was moving into larger facilities but was still done by hand. Many people worked in the place. Cottage industry was disappearing. The new place for making things was known as a "manufactory" from three Latin words, *manus* for "hand," *factus* the past participle of the verb "to make," and *-orium* the suffix for "place where." (This word derivation is like "auditorium," a place where sounds are heard [*audio*].) When the hand was replaced by the machine, the "manus" part of manufactory was dropped and the place became a "factory."

Within the factory, machinery was invented to carry out various operations with less direct input from individual craftsmen. The follower lathe operating on the pantograph principle and the steam engine for power spelled the end of the early type of strictly manual manufacturing. Suddenly, giant machines were needed to manufacture the machines in the new factories. No craftsman, no matter how muscular or skilled, could bore a 12" × 24" cylinder or turn a 12" diameter piston by hand to fit in it. The craftsman had to run machinery. The "factory" was operated by its owner who employed the people within the factory. These operatives still had to be knowledgeable. As skilled masters (master mechanics, for instance), they kept their prerogatives to decide how work was to be accomplished long after they ceased to own the means of manufacture and trade. The apprentice/journeyman/master system adapted itself to the new environment and functioned up until the Second World War. Motion picture training films from that era showed people being trained to do war work on metalworking lathes and the like. Literature on the beginning and growth of manufacturing is plentiful; see, for instance, D. A. Hounshell (1984). His treatise begins, however, well into the era of large manufactories. Improvements within them or at least developments within them are the subject of his writing. He points out that French and American dignitaries right up to Thomas Jefferson were vitally interested for years in producing small arms and artillery that had interchangeable parts. The motivation was the repairability of arms on the field of battle.

After abortive initial attempts by many inventors, a mechanic named John H. Hall made a proposal to the Ordnance Corps of the War Department in 1815 to manufacture 1000 breech-loading rifles of his new design with completely interchangeable parts. Hall told the Secretary of War that he had spared neither pains nor expense in building tools and machinery. He noted, "...only one point now remains to bring the rifles to the utmost perfection, which I shall attempt if the Government contracts with me for the guns to any considerable amount, viz., to make every single part of every gun so much alike that.... if a thousand guns were taken apart & the limbs thrown promiscuously together in one heap they may be taken promiscuously from the heap & will all come right" (Hounshell, 1984, pp. 39–40). If one disentangles the Old English, which makes the manufacture of guns sound like an orgy, recognizing that "limbs" are "parts" and that "promiscuously" means "randomly," then he will see that a gun could be reassembled out of unmarked parts of a thousand guns disassembled and dumped on the floor.

Hall landed the contract for 1000 rifles in 1819. It was a very early example of essentially a Cost-Plus contract. He was given factory space in the Rifle Works, a separate facility at the Harper's Ferry Armory, with the War Department footing all manufacturing cost and with Hall being paid $60 per month plus $1 per completed rifle. The Rifle Works was treated somewhat analogous to the "Skunk Works" at Lockheed Aircraft which turned out the U2 spy plane in the twentieth century. The first set of 1000 rifles with interchangeable parts made entirely with machine tools and precision gauges was completed in 1824. The experiment on random assembly succeeded.

The use of machinery, jigs, and gauges made it possible for laborers rather than craftsmen to turn out the essentially perfect mechanical parts. Use of specialized production machinery made much other high-volume production possible without necessarily achieving identically interchangeable parts. Hounshell (1984) thus analyzes the manufacture of sewing machines, reapers, and clocks. Some of these manufactories did not establish adequate jigs and gauges, and hence got into trouble.

The author has personal experience with one clock that had a manufacturing defect. This ogee mantle clock (circa 1842–1846) still had its paper label which claimed "Warranted Good." However, the manufacturing error had not been fixed under warranty. The symptom of the defect was that the clock would strike 17 o'clock or 23 o'clock or whatever it pleased. Early on, the owners had disconnected the chiming mechanism when they could not fix the chime counter. After getting the clock from his uncle who had purchased it at an estate auction, this author reconnected the chimes and rediscovered the excessive striking. Careful probing along the chime-counting gear with a knife edge showed the manufacturing error. There was a burr on the leading edge of each notch in the rim of the chime-counting gear originating from the cutting of the gear. This gear has a shallow notch, a deep notch, two shallow notches, a deep notch, three shallow notches, etc., up through twelve shallow notches and a deep notch. A finger on the end of a lever slides down a cam surface into the notches one after the other. The finger is ejected from the shallow notches, activating the chime, but is supposed to dwell in the deep notch until the trigger mechanism lifts it out at the next hour. The burr at the bottom of the cam ahead of the deep notch kept the finger from falling into the deep notch, so chiming continued. The author surmises that the burr was the result of a misaligned cutter on an indexed circular table. After the burrs were removed with a fine file by the author, the chime mechanism worked perfectly. The author still has the clock. Presumably all the clock parts were made to be interchangeable, and presumably many clocks chimed 23 o'clock. How many were repaired, how many junked, and how many simply put into the attic is not known.

Interchangeable parts made true mass production possible, as Henry Ford finally insisted. He said, "In mass production, there are no fitters" (Hounshell, 1984, p. 9). Before assembly lines and before true mass production of parts by machine tools with proper jigs and gauges, all manufactories had people in assembly areas who had the title of "fitters." They did the job of "fitting" along with screwing, gluing, riveting, or whatever other assembly method was used. They had to trim, sandpaper, file, or hammer parts until they fit together. It was estimated that 25% of factory effort was in "fitting" prior to mass production. Fitters with rubber mallets were employed to make sheet metal automotive body parts such as hoods and trunk lids (bonnets and boots) fit as late as 1980.

Even some of the manufacturing stage in a handwork shop was fitting. For instance, a dresser drawer had been made of a left side fitted to the front uniquely and then a right side fitted to the drawer front uniquely, each having dovetails measured and sawed by hand. In a mass production setting, a thousand drawer fronts could fit two thousand sides over many weeks. In fact, they could be cut in Connecticut, assembled in Illinois, and sold wherever else the railroads and barges went. Because all interchangeable parts fit, replacement parts became available in one industry after another as the method was adopted. Machine-made circular dovetails for drawers came in about 1870, whereas machines for forming complex gunstocks of wood were invented by 1826 (Hounshell, 1984, p. 38). Progress went by fits and starts.

As far as the philosophy of manufacturing is concerned, the biggest change is not the use of steam, the invention of interchangeable parts, or the introduction of machinery but rather is the array of manpower working for the boss/owner. This new situation spawned the idea of "labor," which had not existed previously. "Labor" being against "management" or "capital" was unheard of in the era of cottage industry, apprentices, journeymen, and independent masters who were shop owners. Along with "labor" came "child labor," the factory as "sweat-shop," wages which were utterly inadequate, "the Company store," profiteering by the owners, and all the other troubles that are a continuing bone of contention between "labor" and "management" with and without "outsourcing." Even without the questions of social consequences of the new concept of "labor," the philosophy of manufacturing had technical consequences that industry is still attempting to rectify. One of these consequences was an inadvertent degradation of quality. The cause-and-effect sequence of this quality debacle will be addressed next.

2.2 Taylor Management Method and Mass Production: Our Twin Nemesis

2.2.1 Taylor's System of Scientific Management

From about 1880 to his retirement in 1911, Frederick Winslow Taylor was a manufacturing theory guru who changed manufacturing and labor in general by introducing time-and-motion studies and new methods of factory organization. He termed his new theory of the management of manufacturing "scientific management." The object of a time-and-motion study was to understand and optimize the way laborers in a factory (or any other region of work such as a construction site) carried out assigned tasks. Scientific management organized all the tasks as well as reorganizing the workplace for efficiency. Beyond 1911, Taylor continued to mentor practitioners in scientific management. His book, *The Principles of Scientific Management,* written near the end of his active career, became the bible of manufacturing organization for two generations (Taylor, 1911/1998).

Taylor developed two sets of principles to govern management and labor in the utilization of scientific management. Taylor observed that management had allowed the old system of apprentices, journeymen, and masters to dominate the new manufacturing job market. Within the workplace, the master still determined how he was to do his work even though he no longer owned the business but was just a laborer. Taylor developed this observation into a philosophy that stated that management had been shirking its portion of the job of running the work establishment. According to this philosophy, management should determine how work was to be carried out and labor should carry out the work. This way, Taylor thought, labor and management would be sharing the work load 50-50 and management would not be shirking (p. 15). The outlook and work Taylor assigned to management falls under four categories:

1. Management should look at the way the master craftsmen did their jobs before scientific management as just a "rule of thumb," which differed from man to man depending on his mentors for

generations back. Management should understand that no "rule of thumb" could possibly be as good as a scientifically derived method for doing the particular job. Management should analyze each job scientifically. The result was to be a "best way" to do the job including written job instructions plus support personnel in addition to the best tools and the best jigs and fixtures to augment the man's efforts. Note later in Chapter 5 how the idea of written job descriptions has propagated forward into the International Standardization Organization (ISO)-9000 quality management standard. There it is assumed that management will have determined the methods before writing them down.

2. Management should use science to select each man for the job. The man no longer had to be a competent master of a trade; in fact, to use modern parlance, many men were "overqualified" for factory jobs. Management should train the man to use the scientific technique of doing the job and supervise him closely if such supervision was determined scientifically, above, to be necessary for the scientific technique to work. The men were to be replaceable since their input except for muscle power was not necessary. In modern parlance, they were expendable. The management was to deliberately wean the laborer away from all the "rules of thumb" with which he was previously imbued as a master by generations of revered mentors. Management was to realize and act upon the realization that in the new factory situation, men could not train or supervise themselves.
3. Management was to cooperate with the laborers in a spirit of hearty conviviality and collegiality to ensure that the men were trained in the scientifically designed work procedures, that the men carried out these procedures, and that the men understood that using the procedures would result in their financial well-being. Part of management's cooperation was to arrange really complicated pay scales so that exceptional workers could earn extra money for exceptional output produced by the scientific method only. Management was to achieve labor peace through this paternalistic outlook and effort.
4. Management was to be diligent in all of the above so that it could feel that it was pulling its weight in the factory, i.e., doing half of the work while the laborers did their half. Management was to do the knowledge-based half of the work while labor was to do the muscle-based half. Taylor thought that the two components of employed persons, management and labor, would then be doing what they were capable of doing.

Note carefully how a laborer's image of himself was sullied and how his years of training and accomplishment were downgraded.

The above is an action-oriented list. Taylor also developed a results-oriented list of four items (p. 74). His scientific management, he thought, would have good results. These might be looked upon as utopian from today's grimmer perspective:

- Inefficient rule-of-thumb eliminated and supplanted by scientific principles.
- Harmony prevails as labor is satisfied and management prospers.
- Cooperation prevails as labor accepts not having individualism.
- Maximized output, bringing prosperity.

Taylor's outlook on laborers has the following foundation. The initial supposition was the observation made by Taylor and others (before he invented scientific management) that laborers deliberately tended to work slowly and lazily. In those days this was called "soldiering" in a derogatory sense, meaning that soldiers did as little as possible as slowly as possible and volunteered for nothing. Taylor ascribes purpose to this behavior, not just laziness (pp. 3–11.) His claim was that the purpose of the laborers was to maximize pay and the number of jobs available to labor. Taylor claimed that the laborers' outlook was that management would lower the pay per part if the parts could be made at a faster rate per hour, thus making the laborers work harder for no extra pay. The variety of paying arrangements available to management in those days was extensive and complex. It is beyond the scope of this book to go into all of them. Let it be said that there was piece work, the day wage, the hourly wage, and a sort of "merit raise" bonus dependent upon the productivity of the individual as perceived by management.

One of these methods open to misinterpretation by labor was the "task management" method of pay (p. 52). A job was set at a certain pay rate per day. A daily level of production was then determined by the scientific management method. This level of production could be met by a laborer working diligently in a sustainable fashion. Time-and-motion studies proved this. If the laborer met or exceeded this level of production, his pay would jump to a higher rate for that day. His pay for the day might depend to the tune of 35% upon making just one more part before the shift whistle blew. The bonus might vary from man to man. In general, the laborer could not be assured that he and the man next to him would be paid the same amount for the same effort, number of hours, calories expended, or any other measure of work, skill, or output. The idea of bonuses was supposed to motivate the laborers to work harder, but perversely in the long run it seems to have had the opposite effect. This is an element of psychology, concerning incentives to generate initiative, which Taylor thought he understood but which modern labor relations experts would say he actually misinterpreted.

In the case cited (p. 52), scientific management was used to increase the output of a factory. Production increased and the laborers were paid more. The ratio, however, was negative as far as the workers were concerned. Production increased much more than 35%. The daily wage of the laborers was raised an average of 35%. The net cost of producing each part went down. In other words, each laborer received less money for producing one part than he had been paid previously even though he got more money overall. In cases like these, it was not intuitively obvious to laborers that they were being treated fairly. Why, they may have thought, should harmony prevail?

Let us further examine the time-and-motion study. The purpose of the time-and-motion study was to scientifically find the answer to a simple question: How long does it take for a man working diligently at a sustainable rate to do the assigned job? First, the job had to be defined. Second, the supplies had to be available. Third, the subjective idea of diligence had to be accepted by both sides. Fourth, the concept of sustainable had to be tested over a reasonable amount of time. The effort had to be expended day-in and day-out. The stop watch was supposed to find the answer. The watcher was also supposed to brainstorm ideas about cutting out useless motions carried on by the worker by force-of-habit.

Time-and-motion studies had been done on animals before. For instance, horsepower had been defined by physicists using horses lifting hay into a barn with a block and tackle. The horses had to work continuously over a protracted period of time to put out a sustainable rate of work without becoming overtired. The result was 550 foot-pounds of work per second. One might surmise that the men Taylor measured felt no better than beasts of burden. For some onerous jobs, Taylor chose men whom he considered to be appropriately "stupid and phlegmatic" (p. 28) like an ox.

Often it was found during a study that getting the supplies or sharpening the tools took time from the defined job. Scientific management proposed to let clerks bring the supplies and let technicians sharpen the tools. Clerks had to be hired and organized. The laborer, especially if he was a master, objected that he should be permitted to take care of his tools and make judgments about how the work should be carried out. Under scientific management, management wrote a job description and instruction sheets to standardize the operation. Ten years as an apprentice, eight years as a journeyman, and many years as a master were superceded by one page of specific instructions. (Note later, in Chapter 5, that instruction sheets are still required by ISO-9000.) The time-and-motion study was often at odds with the culture of the laborer, and the stop watch operator was perceived as an enemy. Management perceived the writing of work instructions as doing its duty, which had been shirked prior to scientific management, since in the old days the master craftsman determined too much of the operations of the plant (p. 10).

The aim of the time-and-motion study was to improve the activities of the laborers. What did "improve" mean? For example, studies were made of loading pig iron onto railroad cars by muscle power alone (pp. 17–31). It was found, very scientifically, that strong men could load prodigious amounts if they were supervised well, told precisely (scientifically) when to rest and for how long, and paid extra. Productivity went up by a factor of four over men simply told to hurry who tired themselves out in short order.

The good pig iron loaders had to be directed as efficiently as the eight oarsmen in a scull are directed by the helmsman in an intercollegiate rowing race.

In other studies, it was found that simple digging was not so simple (pp. 31–35). The best load on a shovel turned out to be 21 pounds. The company had to supply shovels of different scoop sizes for different work materials like grain, coal, iron ore, and sand to permit the right load of material, whether it were slippery or tenacious or visco-elastic, to be picked up and tossed to its destination. The definition of "improve" was to get the most work out of a man in a sustained manner over a protracted period of time. One result is quoted. A savings of $80,000 was affected in a year by 140 men shoveling scientifically whereas 400 to 600 men were required before the implementation of the scientific management task method. The new wage rate was $1.88 per day instead of $1.15 previously.

Still another job studied was bricklaying (pp. 38–41). Initially, the skilled bricklayer got his own bricks, dumped them near his work site, slopped on mortar, leaned down, picked up a brick, positioned it, and tapped it into place with the handle of his trowel. The management man decided that much effort was wasted. The bricks should be brought to the bricklayer in a pack by a laborer and set on a scaffold at a convenient height and orientation for the brick mason to reach and grab. The mortar should be brought to him similarly. The mason should stand on his platform at a good height relative to the wall with his feet toeing out just so. Distances should be arranged so he could reach the bricks, mortar, and wall easily without taking a step. The mortar should be mixed just thin enough so that hand pressure, not tapping, could position the brick. Training showed the man that he could pick up the brick with one hand and spread on the mortar with the other, saving motions and time. The essence of the improvements was to eliminate unnecessary motions, provide mechanical aids, and teach motion economy. The result of implementing the scientific method was an output of 350 bricks per hour, up from 120. Some foreign unions at the time were limiting their workers' output to 375 bricks per day by comparison.

In a more general sense in a production environment, the "improvement" was carried out to eliminate inefficiencies in the way the laborer moved and the way work flowed past him so that the laborer could finish one part and move on to the next (identical) part in as short a time as possible. The entire operation of a factory was reorganized by Taylor for "efficiency." Each man as well as the entire shop was supposed to operate at the highest possible efficiency. The purpose in this was to get the most production per hour out of a laborer and the machine he worked at. (One wanted to eliminate extra machines because it was Taylor's belief that American industry was overfacilitized wastefully at the time.) Taylor desired every laborer to be moving every relevant part of his body constructively at almost all times. Downtime was allowed (actually enforced), as Taylor discovered that rest was necessary to promote maximum efficiency. During work, one could generalize the following scenario: the right hand was to be pushing the partially assembled object to the next man's station on his right while the left hand was picking up a screw to insert into the next object pushed over by the man on his left. This idea came into being before the invention of the moving production line but fit in perfectly with Ford's to-be-developed new moving production line. Wasted motions were to be engineered out of the system. The operation of management and labor together was supposed to be smooth, harmonious, and cooperative (see results-oriented Four Points outlined earlier). It is not clear why Taylor thought that the laborer would feel in harmony with the manager who was stripping his *modus vivendi*, namely his prerogatives as a master mechanic, from him. It is necessary to examine what the laborer thought.

The actual operation of the time-and-motion study must be examined. Taylor's instruments of research were a pad, a pencil, and a stop watch. As he approached a man on a production line, the laborer soon intuited the fact that he would be expected to work harder and faster. One disciple of Taylor's in a major report on an application of Taylor's method (Parkhurst, 1917, pp. 4–5) mentions that rumors about the expected "hustling" would precede the approach of the efficiency expert by days and produce a bad rapport between the expert and the laborers. Parkhurst avers that forcing laborers to hustle was not the aim; rather, eliminating inefficiencies was. In theory, Taylor was benign and altruistic, seeking only to eliminate waste. Wasn't waste an ignoble thing, and wasn't the elimination of waste a fitting way for an intelligent man to use his career? Waste not, want not. One cannot fault Taylor *a priori* on this valiant attempt at morality.

To accomplish the banishing of waste, Taylor had to develop a corollary to his practice of time and motion. He realized that every man had to be interchangeable with every other man on the job. Even though the individual man might be trained, coached, supervised, and paid specially, almost any other man could take his place. All the jobs had to be reduced to such a simple level that anyone hired "off the street" could do any of them. These hard-and-fast rules had exceptions. Taylor realized that some people simply would not fit some jobs. He also discovered by experimentation that management needed to do a major amount of planning, training, and supervision. This hiring methodology was already the practice of management, so Taylor's theories fit in perfectly with existing management regimen. All the intellectual content, skill content, and thinking requirements had to be banished from every job. Taylor felt that accumulating all the knowledge of all the craft masters into the annals of the company's management under scientific management let management perform its responsibilities while it had relied upon workers too much in previous times. Laborers were left with manual tasks for which they were better suited. Taylor thought that management would be pulling its weight more equally in the management-labor team under his system (Taylor, 1911, pp. 15–16). A minimal amount of instruction from the newly omniscient management permitted a laborer to do the small finite number of motions in any job. The job content had to be completely described in writing. All the thinking had to be "kicked upstairs." Each level of supervision from foreman to first-line supervision to middle management on up had to do the minimal amount of thinking to accomplish interchangeable jobs among which men at the particular level would be interchanged. The person at the next higher level of management would have the responsibility to think about anything slightly unusual, like a production problem. Even the capability of recognizing a problem was deemed unnecessary on the part of the production worker. Once the factory owner and a small coterie of experts with unquestioned power (Parkhurst, 1917, p. 4) had built and equipped the factory and established its procedures, no intelligence ever need roam its halls again. All the planning and thinking was up in the Planning Department, the Scheduling Department, and among other management functionaries. This is part of the legacy Taylor left to modern manufacturing

At the same time, Taylor was searching for inefficiencies in the way the factory owner and his small coterie of experts might have organized the work of whole departments and divisions of their company, not just the inefficiencies of the work of individual men. Gross organizational inefficiencies were discovered in many companies. Parkhurst (1917) reports one set of inefficiencies in a machine tool company he consulted for early on. Optimizing company-wide organization is the other part of the legacy Taylor left. His type of organization is the type providing the barriers that W. E. Deming wishes to break through and eliminate (see Chapter 4). Much of what Taylor accomplished would now be termed suboptimization of the company by optimizing separate segments of it.

Parkhurst's book (1917) reports his success at this interchangeability of people in reorganizing a company of about 100 employees along the lines of Taylor's theory (p. 2). These 100 employees had been operating in a milieu somewhat disorganized with an efficiency below 40% (according to Taylor's method of calculation carried out by Parkhurst). After the reorganization of the company according to scientific management with changes in job descriptions, departmental lines of reporting, etc., the same 100 laborers could perform all the new jobs except one. After trying out all 100 laborers at this job over a period of two years, Parkhurst found that he needed to hire someone from out of town with extra skills to do this job. Parkhurst attributes the improvement in efficiency, plateauing at 90% (again, his calculation of "efficiency," which he does not define) to be due to the new scientific management system and only slightly to the one new employee in 100.

In the machine tool company, Parkhurst (1917) achieved substantial cost savings with his application of Taylor's method. One table shows the labor time and resultant cost reductions in the manufacture of 275 parts used in various models of punch presses. As bonuses were introduced for some workers and not others, the relationship is not exactly linear. Labor times per part were reduced by 30% to 80% or more. Inventory control was improved, making final assembly of the machine tools more efficient and improving the rate of filling orders.

If one did not have human psychology to deal with, the initial successes of the Taylor method would have been easier to sustain. If human beings acted as robots and if root causes of failures never occurred, then Taylor management would have worked as desired. On these two problems, it must be said that Taylor created the first and overlooked the second. Human beings have foibles, reactions, and pride. They like to have inputs to situations including production. The idea of "kicking all decisions upstairs" is contrary to the average laborer's pride. Witness the more modern Japanese idea of Quality Circles and the European idea of assembly stations with teams introduced in the 1980s. (Interestingly, Ford cars before the Model T were assembled at assembly stations where piles of parts were added to a chassis by a team of workers [Hounshell, 1984, p. 220].) Workers in these modes organize their own work somewhat and get to the root causes of problems. To visualize how much human capability is wasted by "kicking the decisions upstairs," think of all the "do-it-yourself" activity these laborers plan and carry on at home after their shifts.

Other writers have addressed the consequences of the Taylor philosophy. One in particular, M. Walton (1986a, p. 9) while concentrating on Deming and his philosophy, stated some background on Taylor. Her analysis differs little from the material given above. One interesting factor she notes is that much of the labor affected by the Taylor scientific management method was uneducated immigrants arriving by the boatload before the reactionary immigration laws of the 1920s. These people, all in need of jobs, could be interchanged at will by management. Walton does not mention the highly trained American masters and journeymen who were disenchanted by having their knowledge base debased as "rules of thumb" when they went to work in factories. It should also be emphasized that Taylor invented scientific management, practiced its implementation, and retired as an active implementer before the moving production line was invented at Ford's.

The idea of the "efficiency expert" with time-and-motion studies and ideas about all sorts of waste management has been treated even as comic. A semibiographical book and film on Taylor's life and times were *Cheaper by the Dozen* (F. G. Gilbreth, Jr. and E. G. Carey [1948] and Twentieth Century Fox Films [1952]). This comedy portrayed big families as efficient because of older children taking care of younger ones and because of the availability of hand-me-downs. A truly hilarious scene shows Taylor demonstrating the most efficient way for one to soap himself in a bath. However, the fear engendered at the work station by the approach of the efficiency expert is not addressed.

Next it is necessary to study the logical culmination of Taylor's interchangeability of men along with the Hall's "promiscuous" interchangeability of parts in Ford's mass production philosophy.

2.2.2 Ford's Extensions and Changes

The scientific management method did not lead directly into the Ford moving production line. People tend to think of the efficiency expert as interacting with the person working on the line to improve his performance. This is actually far from the case. It is necessary to study the Ford system of mass production to see the changes and to understand the culmination of the difficulties that Taylor initiated and that the moving production line exacerbated.

Taylor and his followers had been organizing every sort of enterprise starting in the 1880s. It would be logical to assume that the methods of scientific management found their way into the fledgling automobile industry, as Taylor had worked considerably for the metalworking industries (Taylor, 1911/1998, pp. 50–59). In his chapter, "The Ford Motor Company & the Rise of Mass Production in America," Hounshell does not mention scientific management until he is 32 pages into the description of Ford's operations (see Hounshell, 1984, pp. 249–253). The initial mention of scientific management is enlightening. It is reported that Taylor gave a speech to a management gathering in Detroit in which he claimed that the automobile industry was quite successful at introducing scientific management in its workplaces. Taylor went on to say that the industrialists had succeeded on their own without hiring expert consultants employing Taylor's formulation of scientific management. Some industrialists disagreed to the effect that they had actually anticipated Taylor's method earlier on their own. It is reported that Henry Ford claimed that he developed his manufacturing system without recourse to any formal theory.

However, reading of the Hounshell chapter (1984) will show that the young mechanics whom Henry Ford hired to design his factories and automobiles were using the generalized principles of scientific management intuitively for factory layout just as Parkhurst (1917) had done formally as an expert in scientific management. This observation about Ford engineers was true right up until the introduction of the moving assembly line. Then everything changed. Prior to the moving assembly line, in their Planning Room Ford's engineers were laying out factory plans on "layout boards" (Hounshell, 1984, pp. 228–229) with moveable cutouts to represent each machine tool. These two-dimensional miniatures allowed them to plan the placement of the tools sequentially in the order of the work to be done on each part, so that a part could pass from one machine to the next with the minimum of logistics. Parts being manufactured in more than one operation were to be treated by a sequence of machines arranged thus. Even heat treating furnaces were placed sequentially among the machines. One no longer had to go to the Lathe Room to turn Part X on a lathe and then to the Press Room to punch one end of it flat. Machines in the Part X Manufacturing Room were arranged in order of use. Ford introduced what Taylor would have termed overfacilitizing with special-purpose machines in order to turn out identical parts at a much faster rate than Taylor had dreamed of.

Ford envisioned manufacturing great numbers of vehicles that were to be inexpensive, rugged, easily repairable, lightweight, and simple to operate. Some wags said "FORD" meant "Fix Or Repair Daily." Basically Ford needed mass production and was the one person to whom this technique owes its realization. His goal was to sell huge numbers of automobiles to the general public. He recognized that he needed interchangeable parts and efficient production as well as a viable design. His consummate early design, the Model T, was the result of work of brilliant people he hired for both design and manufacturing functions (see Hounshell, 1984, pp. 218–229). For the Model T, he built factories and special-purpose machines that not only produced efficiency and accuracy, but that also could not be adapted to build a new design when the Model T finally became obsolete. But that came 15 years later.

In essence, all the ideas pioneered by Taylor about organizing a firm in departments and organizing production in factory situations with minimum wasted motion such as logistics were incorporated into the Ford factories. Taylor was improved upon considerably, one can see, by following the account in Hounshell (1984). Other industrialists were doing the same. All that was left to be invented was the moving production line. Team assembly was being done until that development (Hounshell, 1984, p. 220).

Hounshell (1984, pp. 217, 241) points out the surprising fact that the idea for the moving production line arose from the moving "disassembly lines" for carcasses in the Chicago slaughterhouses. A dead steer hanging by its rear hoofs would slowly and systematically disappear until some remaining bones were shipped to the glue factory. The inverse idea did not jump out immediately. It took Ford's initiative and motivation to produce many automobiles rapidly until 1913 to debut the first moving assembly line. Then, like the slaughterhouse in reverse, the automobile came into being from "bare bones" of a frame until, some hours later, it was gassed up and started and driven off the assembly line, finally "alive." Other examples of moving sequential production lines available to the auto engineers for study and inspiration were in flour milling, beer making, and food canning.

The moving assembly line was created at the new Highland Park factory specifically for one mechanical subassembly of the Model T and started up on April 1, 1913. This subassembly was the ignition magneto mounted on the flywheel. The parts, dragged along by a chain, were at waist-height on a slide with the men standing alongside it and screwing in components. The new assembly system allowed the work force to be reduced from 29 to 14 while reducing the assembly time from 20 man-minutes to 5 man-minutes per subassembly (Hounshell, 1984, pp. 247–248). The success of this moving assembly line was met with jubilation in the company and motivated the initiation of experiments on moving assembly lines for many other subassemblies. By November 1913, after experimentation, engines on a duplex moving line were assembled in 226 man-minutes instead of the previous team expenditure of 594 man-minutes.

By August of the same year, a moving assembly line was in the experimental stage for the final assembly of a vehicle chassis. (In those days the body was added on top of the chassis later.) While crude, this line made great gains in productivity and pointed the direction for further development. At least five major iterations of complete redesign accompanied by many improvements, particularly in delivery of subassemblies to the line, are listed by Hounshell (1984, pp. 253–256). Development was so rapid that by April 30, 1914, three essentially complete automobile assembly lines were in operation turning out 1,212 cars in eight hours. The actual effort expended per car was 93 man-minutes from hooking a bare frame to the line until the chassis was finished. This contrasts with the previously required 12.5 man-hours with the static team assembly method. Mathematically, this represents an increase in productivity of a factor of almost 8. The chassis line is what is remembered by the general public as the first moving production line although the subassembly lines preceded it and were prerequisites for the acceptance of the idea even as an experimental entity.

The Model T had been manufactured at other factories since its roll-out on March 19, 1908. Production on the car, already a wild success by 1912, almost tripled in 1913 and climbed to almost 600,000 in 1916 (Hounshell, 1984, pp. 219–224). With the advent of the moving assembly line, all the ideas about static assembly such as team assembly were scrapped.

The Ford staff initiated many innovations. The "best and the brightest" at Ford's were so sure of their new production methods, jigs, fixtures, and measurements that they had the audacity to assemble an engine into a car without ever running the engine. The first time the engine was started was at the end of the line as the car was driven to the lot to await loading onto a train for shipment. The engineering management maintained that accurate manufacturing would make everything turn out correctly in the end.

Let us now look at the differences between Taylor's scientific management and Ford's mass production with respect to the worker (see Hounshell, 1984, pp. 251–259). We have already ascertained that efficiency in factory layout was a goal of both and was achieved by both in their own sphere of activity. What differences affected the worker, and what were the results?

Taylor made the underlying assumption that the job was defined *a priori* and that science was to be applied to maximize the efficiency of the laborer doing the job. The maximum output was to be obtained from each worker at a preexistent job by time-and-motion optimization, by training, by supervision, and by bonus pay. In the Taylor system, the man's getting to the job was expedited optimally.

Ford, on the other hand, made the opposite assumption. The job was not defined. The job was to be invented. This was to be done by inventing a special-purpose machine to do the job and placing a man next to the machine to do minor functions. Other jobs were to be invented which consisted of assembling things made by these new machines. The Taylor ideal of interchangeable men was brought to fruition. The men, however, did not move from here to there but were essentially stationary.

While Taylor wanted the men to work fast but sustainably, Ford wanted the men to perform the machine's minor operations at a speed the machine dictated.

The rapid worker was to be slowed down simply because the machine or moving line did not go faster, and the slow worker was to be speeded up to keep up with the machine. Time-and-motion studies determined the rate at which a machine or a line ought to operate with workers not permitted to slack off.

Under mass production, the men no longer needed any training or any skill to do their work. They did not need to be taught the optimal way that the best master craftsman did the job in order to emulate it as in scientific management. The man became an appendage of the machine.

Serious labor problems ensued. Labor turnover rose to 380% per year, an unheard-of number. Ford introduced the wage of $5 per day to entice the laborers to stay on and "marry" themselves to their machines. People felt themselves to be selling out to voluntary servitude. As one harried wife wrote, anonymously, to Henry Ford about the moving assembly line, "The chain system you have is a *slave driver! My God!*, Mr. Ford. My husband has come home & thrown himself down & won't eat his supper—so done out! Can't it be remedied?… That $5 day is a blessing—a bigger one than you know but *oh* how they earn it" (Hounshell, 1984, p. 259). The idea of labor unions having a say in work rules and line speeds did not come to fruition for 20 years or more. The Taylor ideal had finally come to pass. All knowledge was kicked upstairs. All men were interchangeable. One did not even have to be strong to shovel or carry. Indeed, Taylor was surpassed and bypassed.

2.2.3 Further Notes on Taylor and Ford

The idea of "hiring off the street" was used by Taylor and by the mass production philosophy that followed him and grew out of his work and the work on standardizing interchangeable parts. "Hiring off the street" was corroborated by some oral history I was told by a gentleman (Papadakis, 1975) who had been working his way through college as a young man in the 1920s. This was at a point in time when Taylor's methods had become second nature to industry and when industry was moving forward to new approaches to making the laborer even more of a cog in a great machine. The young man and a large crowd of men were standing outside the gate of a Detroit factory that had advertised for workers. A company representative appeared and yelled that they needed punch press operators. The young man turned to the man next to him and commiserated that he really needed a job but had never operated a punch press. The buddy in line with more "street smarts" told the young man to go up to the company representative and say he was a punch press operator. The buddy continued with the instruction to follow the foreman into the factory and when presented with a particular machine to operate, just say that he needed instructions because he had never seen that particular type of punch press before. The strategy worked. The young man got the job.

An example of the waste brought on by Taylor management theory as carried forward into modern factory practice is given in the following. This is a short report written by the author (Papadakis, 2001, 479–480) in an NDT professional magazine and reproduced here. The factory problem reported on what happened in the 1980s. This shows how pervasive and invasive the influence of Taylor has been. He did not go away even when we knew he should have.

> It was an emergency. Another emergency. When you're up-to-there in alligators you can't hardly drain the swamp, but we had to catch this alligator in a hurry.
>
> Transmission Division came to us with a problem with spot welds. They knew we had developed an ultrasonic method to test spot welds and they called us immediately to solve their problem.
>
> Spot welds were being used to hold certain brackets to the interior of the steel cases of torque converters for automotive transmissions. Torque converters take the power of the engine and transmit it through an impeller and a turbine combination to the gears and bands in the automatic transmission so that the power can get smoothly to the drive wheels of the automobile or truck. The torque converter case is 2 pieces of sheet metal which come together in the shape of a bagel, about 14 inches in diameter, which has been sliced and put back together. Continuing the analogy, all the interior dough has been hollowed out to permit the insertion of the impeller and the turbine.
>
> At any rate, the spot welds on this bracket inside one side of the converter case were failing. No amount of adjusting current and voltage could produce good welds on this new model of torque converter.
>
> Using his previous work (Mansour, 1988) based on even earlier work at the Budd Company in Philadelphia, Tony found that he could test these spot welds and predict future failures. Based on this success in the technical feasibility study on a few converter cases, Tony and I were invited to visit the transmission plant and recommend a manufacturing feasibility study and then suggest automated implementation equipment, namely a big, expensive system.
>
> After talking with the very worried engineers and their harried managers, Tony and I were taken into the plant to observe the spot welding equipment in action. The equipment was massive and heavy-duty; running into it with a Hi-Lo couldn't damage it. There were two spot welding heads mounted on a large piston running vertically which brought the heads down to touch and clamp the bracket to the section of the converter case. These two heads were placed symmetrically at 3 o'clock and at 9 o'clock with respect to the shaft hole in the center of the circular section of this half of the converter case "bagel."
>
> The end of the piston was insulated from the rest of the machine so the current when introduced into the spot welding heads at the bottom of the piston would flow through the welding heads, through the two layers of sheet metal to be spot welded, and into the corresponding

lower spot weld heads. A region of metal was supposed to melt where the current passed from one metal sheet to the other, and then refreeze into a nugget when the current was turned off.

The metal parts were placed into the jigs correctly, the piston moved up and down correctly, and the current flowed.

The current was so many thousands of amperes that it had to be carried by large amounts of copper. Because the current source was stationary and the piston tip required the current to move up and down with a sizable throw, the cable for the current had to be flexible. For symmetry to the two spot welding heads, the current was brought to the piston head by two sets of conductors from the two sides (see Figure 2.1). To be flexible and to have a large surface

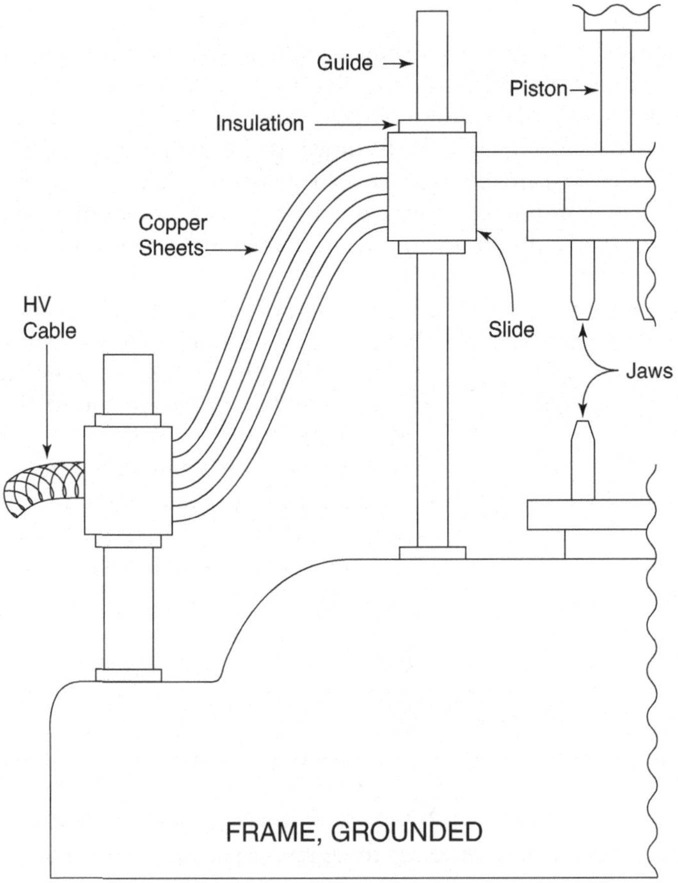

FIGURE 2.1
The welding machine with its jaws open (up position) before the insertion of the part to be welded. The current is to be carried by the curved thin sheets of copper drooping next to the electrode connections.

area to carry the large AC current, the conductors were multiple layers of thin sheets of copper separated by air gaps of about ten times their thickness. These sheets of copper were clamped to the piston at the center and to two electrical buses outboard. To permit the flexibility for the vertical motion of the piston, the copper sheets were extra long and drooped down in a sort of catenary shape on the two sides of the piston. As the piston raised and lowered, one could imagine looking at the cables of a suspension bridge flex if one pier moved up and down.

Tony and I watched this welding process intently. Soon the NDT solution became obvious. What was happening was this: With the pistons in the lowered position, the bottom copper catenary sheet was touching the frame of the machine and grounding out! (see Figure 2.2).

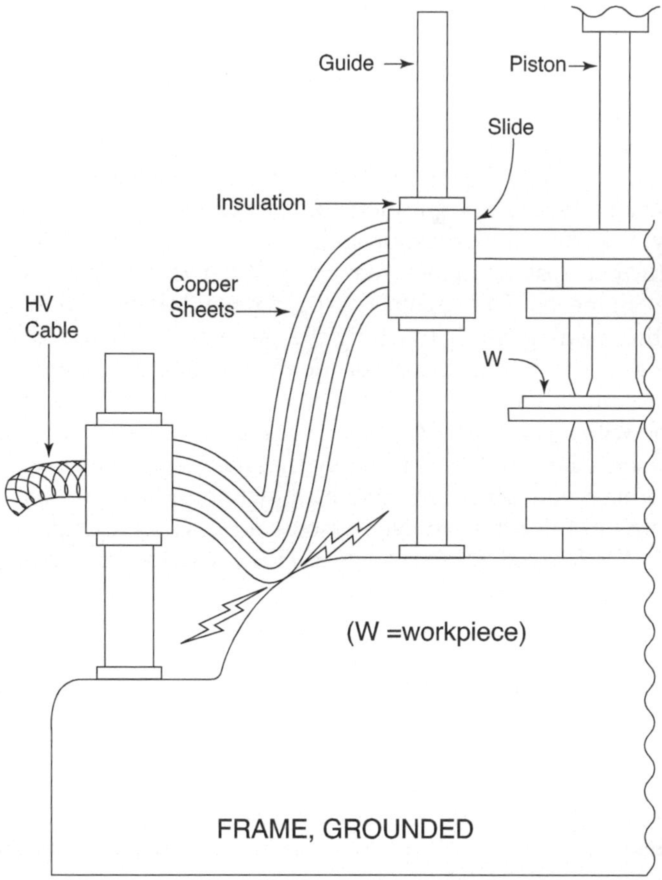

FIGURE 2.2
The welding machine with its jaws closed. The length of the welding heads was short enough to permit the copper sheets to short out on the frame of the machine, causing inadequate spot welds.

Occasionally (but not on every stroke) sparks would fly as the current came on.

The NDT Solution was to do no NDT at all. First, we ordered an electrician to tape large pieces of 1/2-inch-thick rubber pad to the frame of the machine under the copper catenary conductors. Second, we recommended that the engineers adjust the heights of the various elements so that the copper would not droop so far, leaving the copper catenaries up in the air where they belonged. Good spot welds resulted when the current was subsequently set to its design specifications.

The root problem here was the old-fashioned Taylor theory of management which was in use in all of American industry for so many years (see, for instance, Walton, 1986a). Under the Taylor regimen, the workers on the floor just moved things and had no intellectual input into a process. Even if the workers had reported sparks in the wrong place, they would not have been listened to by the foremen and would not have been believed. Indeed, they would have been reprimanded for interfering and not producing. Taylor did not want workers to have any training, so they would not have even been instructed that sparks or electrical lines grounding out were undesirable. The engineers would not have been empowered to ask the workers any questions of substance.

Taylor kicked all the responsibility upstairs to the engineers and then further upstairs to their managers. What was the responsibility of the engineers? The engineers would have drawn up the process with the welding heads touching the piece parts in the right place and the current being correct for the thickness of metal, and would have assumed that their spot welder would work well in a turn-key fashion with its manufacturer being the responsible party ("upstairs" from them). It would be very likely that the engineers' analysis did **not** go deep enough (before the introduction of Ishikawa fish-bone diagrams in the 1980s) to even discuss a possible short-circuit in a Failure Modes and Effects Analysis. Who would have thought it, anyway? But even in the presence of acknowledged difficulties, the engineers did not have the time or did not take the time to go to the plant floor and look at what was really happening. And, under Taylor, the managers were absorbing blame but doing nothing intellectually creative. The alligators were taking over the factory as well as the swamp.

Taylor was alive and well and living in Dearborn as well as in every other American manufacturing city. Kicking thinking upstairs had been initiated by Taylor and completed by Ford in mass production.

2.3 Quality Degradation under Taylor Management

And what is the real Taylor legacy? Besides the fear (perception) every laborer developed of being forced to work faster and harder at no increase in wages (even though Taylor did pay extra), the laborers lost control of the quality of their output. This condition was made even more severe by mass production. The foreman wanted output of a certain number of parts per shift. As the laborer was paid by piece work or else given his bonus only if enough parts were produced (task management), he would get docked or fired for not producing this amount. On a moving line, some of management's control was lost. As one laborer passed a part along to the next fellow down the line, the next fellow never had time to ascertain whether the one before him had finished his operation. In fact, it was a matter of honor not to question your buddy's work because both people wanted to produce the maximum number of parts without interruption. Labor as well as management wanted a large amount of output.

Other systems had analogous theories of output from labor. Americans in the 1950s and later were accustomed to criticize the Soviet system of "norms," which required a certain number of parts per day per worker. The American system was no more just. Even when the American system changed from piece work to hourly labor, the norm was still there as determined by the line speed of the production line. Even today the production line is virtually the same. In a modern advertising brochure, the Ford Motor Company reported that line speed in the Rouge Assembly Plant assembling F-150 trucks is 67 units per hour (Ford Motor Co., 2005). While the old concept of line speed is still operative, certain ergonomic improvements for the laborers are also reported. Back when production lines were invented, the line speed was determined by the efficiency expert, so the variability of the labor force could not be taken into account. If you had a "bad hair day" in the old days, or worse, you could get fired. If a laborer realized a problem in his work station and its required rate of output, he could not complain or make a suggestion for fear of being fired and replaced by a stoic individual who could not or would not think. Actually, refusal to think was a defense mechanism for job security.

Of course, some thought went into slowing down production by deceiving the efficiency expert when he evaluated a work station. To be able to work slower had advantages such as safety as well as quality. The laborers could recognize the value of slowness while it was despised by the efficiency experts as sloth. For instance, in the matter of safety, a punch press could take off a hand if you were in a hurry to activate the press while still positioning the part in the jaws. To increase speed, the efficiency experts designed the presses to be loaded with two hands and activated by a foot treadle, a sure formula for tragedy. Not until the Occupational Safety and Health Administration (OSHA) in the 1970s did the engineering modification of two-handed switches to activate dangerous equipment become a requirement. Having time to finish the operations at your work station even if you had to blow your nose obviously would have increased overall quality but was not allowed by the efficiency experts. It could be argued that the laborers were not motivated by quality concerns or even by safety per se but by the quality of life in the workplace. The effort of management to increase the tempo of work was seen as a deliberate effort directed against the lives of laborers. Labor retaliated by not showing concern for the goals of management. Just doing the minimum that management wanted was adequate; one does not have to imagine sabotage or any criminal activity.

Jealousy compounded the question of work ethics. In the Taylor system, to increase the rate of production, some employees but not others were offered bonuses to produce more per hour than the required amount. Parkhurst (1917, p. 7) treats the bonuses as a valid and valuable motivating tool under scientific management which allowed the good laborers to advance in income to their maximum competence. Laborers, on the other hand, all wanted to be paid a day's wages for a day's work. Many years later the arrangement became a negotiated contract.

Management itself contributed to the degradation of quality by wanting to ship as much material as possible out the door. Only when the parts fell apart as in the torque converters mentioned above did the management pay any heed to quality. The worse situation was when poor quality resulted in returns of unacceptable material by good customers. A motel owner once told this author that 31 new desk-and-chair sets had been ordered for the motel. A total of 26 sets had to be returned with faulty glue joints. This is the typical result of hurrying to produce and ship.

In the days of the journeyman as craftsman and the Master as responsible entity, the joints would have been done right the first time. Absent this motivation and capability in a Taylor or mass production arrangement, what was the approach tried next?

2.4 The Inspector as the Methodology To Rectify Quality

As the failure rate of the production system approached 100% so that nothing could be shipped without being returned, the management of factories where the deterioration was severe realized by simply looking at the balance sheet that some remedial action had to be taken. Put succinctly, bankruptcy was just around the corner.

The first knee-jerk reaction was to introduce end-of-line inspectors who would reject faulty production so that it would not be shipped. At least its presence would no longer be an embarrassment in the marketplace. If caught, the faulty items might be repaired if any potential value were left, and later shipped and sold. This is the great bug-a-boo of "rework."

A production line might settle down to a production of 100, shipment of 75, repair of 20, junking of 5, and still experience a return of 5 of the first 75 shipped because of the inherent imperfection of inspection (before high-tech means became available.) Inspection was always acknowledged to be imperfect in this sense. Sometimes double inspection, organized sequentially, was instituted to catch the missed nonconforming parts. Some investigators have called this counterproductive, claiming that the first-level inspection simply became lazy. Of course, no matter how many repetitive inspections were performed, latent flaws and improper intrinsic properties of materials could not be detected by inspectors. Spark tests for hardness, for instance, were only visually approximate. Many laboratory tests were destructive and could not be performed on all of production. Some required special test pieces that were incompatible with mass production, causing intolerable interruptions if made on-line or not being representative if made off-line.

Flaws or errors deeply embedded in final assemblies made it necessary to embed more inspectors along the production line to catch errors earlier. It has been reported that as much as 26% of the labor force in some automobile factories was composed of inspectors even in the recent past. These inspectors certainly rejected much nonconforming material. If the regular laborers had "done it right the first time," the inspectors would have been unnecessary except for latent flaws, which could not have been detected anyway. Besides the cost of their wages and the value of the space they occupied, there were other untoward effects of the presence of inspectors.

The first problem arises from the assumption that the other laborers could have "done it right the first time." Could they have? After all, management had organized the work effort by Taylor's method or by the mass production moving assembly line method which both forced errors upon men. Was the degradation of quality labor's fault, or management's fault? Deming lays the blame at the feet of management (Deming, 1982). Deming's ideas will be explored further in Chapter 4.

The second problem is a corollary of the first and can be characterized as finger-pointing. Assuming that the laborers were wrong, the inspectors (also laborers) were blaming the production laborers for the poor quality made inevitable by management. Nobody knew it was management's fault until decades later. Hence the finger-pointing became bitter adversarial behavior involving foremen and so on. In his report on the application of the Taylor method, Parkhurst reports in at least two places (Parkhurst, 1917, pp. 61, 128) that the immediate action of a manager upon finding an anomaly is to assign blame. Investigating the root cause of the difficulty was not even considered. Assigning blame is now known to be counterproductive and improper psychologically. Upon discovering a problem, the inspector and the laborers plus a management representative should search for a "root cause" rather than assign blame. The counterproductive finger-pointing leads to the next question addressed in the next section.

2.5 Adversarial Confrontation: Inspector as Cop and Laborer as Crook

How did the production laborer perceive the inspector? From the point of view of the laborer on the line or in the shop, the inspector was a policeman. The cop made it his business to find something wrong with production even if the laborer had no control of the process. Thus, the inspectors developed an adversarial position relative to the production laborers. As the job of the inspector was to find errors, and as he would be criticized by management if he did not find errors, the inspector began finding errors where there were none, exacerbating the situation. With the situation hopeless, the line laborer developed a "don't care" attitude. The laborer simply wanted to get enough parts past the inspector (passed by the inspector) to get paid for his piecework or quota. The production laborer certainly did not want to be called onto the carpet for poor performance, which might cost him his job. In the mean time, the inspector continued on his mission to find errors to keep his own job. This adversarial situation with the inspector as a policeman and with the laborer as a crook (exacerbated by the policeman's acting unjustly from the point of view of the laborer) led to the same scofflaw behavior that was happening in society as a whole. The then-concurrent situation in society was Prohibition, where the cops were perceived as persecuting ordinary citizens for exercising their natural right to take a drink. Finally in the early 1930s Prohibition was repealed. (As an aside, it is interesting in this legalistic context that the Constitutional amendment creating Prohibition was declared unconstitutional by another Constitutional amendment.) However, back in the factory the legalistic charade went on. The controlling situation of the inspector and the Quality Department that evolved was not discontinued. Labor remained an adversary of management in the quality realm. So, we must ask, what long-term effect did the inspector have on the realm of quality?

2.6 Ineffectuality of Inspector To Improve Quality

It can be asserted unequivocally that the inspector impacted quality. But how? The inspectors may have raised the outgoing quality in shipments from the factory because they caught a certain percent—possibly even a large percent—of the defective items produced. The inspector never detected every bad part. However, the production had to be proportionately larger than the norm in order to ship as many as planned. But then, the extra production had to be inspected, resulting in some of it failing. Then, even more had to be produced to meet the shipping requirements, and so on.

Quality as-produced did not improve except by accident. One never knew when a similar accidental occurrence might send quality plummeting to a new low. Production was increased even further to provide a backup in case there was not enough good production in a time period to ship. One called this "Just-in-Case" inventory. All this extra production and inventory incurred costs not only of the value of materials and the wages to pay men, but also interest on bank loans to float the inventory and so on.

This author has worked with individuals who attest to the idea that, without the inspectors' knowledge, faulty production was sequestered for later shipment at a time when actual production could not meet the demand (Kovacs, 1980). The foreman wanted material to ship, and the management was bypassed. Or perhaps the management wanted material to ship. The inspector, for all the ire he raised, did not manage to raise quality itself. Rework and extra production were always the norm.

Was there any way out of this morass? Management tried one way to attack one aspect of the problem. That was perfecting the inspection process by electronics. The principal idea was to eliminate finger-pointing by objective, true measurements. The next section addresses the approach taken.

2.7 The "Perfect" Inspector: Automated 100% Inspection by Electronics

As electricity progressed to electronics and new techniques burgeoned especially after sonar and radar, electronic methods of inspection were invented. Management opted to improve inspection by electronic means. Electronics promised to detect essentially 100% of nonconformities. Beyond that, it promised to detect latent defects and intrinsic physical properties previously inaccessible.

This section will speak of inspection by electronic means in a generic sense. A whole chapter is reserved later in the book (Chapter 8) for the discussion of particular methods and instruments. Suffice it to say that the "electronic means" include AC electrical induction, DC currents, audio sounds, x-rays, ultrasonics, atomic physics, nuclear methods, isotopes, optics, infrared, and many others.

These systems with their sensors are interfaced with other electronic circuits to make YES/NO or GO/NO-GO decisions when the sensors encounter nonconforming material. These decision circuits activate different paths for the good and bad material to traverse. Rejects are carted away automatically.

These systems are characterized by being rapid and accurate. The accuracy is characterized by a Probability of Detection, which indicates a trade-off between Type 1 and Type 2 errors. One can make the detectability of faulty material almost as high as one would like by accepting the scrapping of a few good parts on the borderline. Latent defects and intrinsic physical variables of many kinds can be detected electronically. Thus, the electronics is more than just a substitute for the manual observation. Further discussion will be given in Chapter 8.

As more and more types of electronic systems became available, management bought and installed them to ensure that poor quality did not get shipped beyond the point in the line where they were installed. A lot of systems were for outgoing inspection and many more were for Verification-in-Process as it is now called.

Where it was more cost-effective for a laborer to manipulate the probe or place a part near a probe than to pay for automated materials handling systems, then hybrid man-machine systems were installed. Every citizen has seen some hybrid man-machine systems involving electronics. Bar-code readers in modern retail stores are an example of a hybrid system. Thermometers a nurse sticks in your ear are another.

Management bought and installed industrial systems with good intentions but without the complete understanding of the way they should interact with quality itself. As with human inspectors, the assumption in the 1940s up through the 1980s was that the purpose was to install a "perfect" inspector to make sure no faulty material was shipped or placed further into production. Feedback to cause better production in the future was not a consideration. Whether it would have been possible or not at an early date is another question. The effort and/or imagination to create a synthesis between automated 100% inspection and Statistical Process Control to stop a process when it went out of control and began producing nonconforming parts did not come about until 1985. This epiphany event of invention will be addressed in Chapter 3 on SPC and Chapter 8 on NDT. In the meantime, it is valuable to address the correct attitude toward 100% inspection in generic terms.

2.8 Fallacies of Early Implementation of 100% Inspection

One would want to pinpoint any fallacies in the logic that has led up to the installation of 100% inspection of parts to preclude future illogical behavior. The question arises whether it is possible for management to find situations in which to install 100% inspection is an "open and shut case" in the affirmative. One definite positive case can be characterized by the following example:

1. The factory needs raw material without cracks.
2. No supplier can sell us perfect material.
3. A test can find the raw material sections with cracks and discard them.
4. The good areas are large enough to make our parts out of.
5. Therefore, install the test.

Examples abound. Among them are heavy wire to be headed into valves, rod stock to be pierced for PGM tubes, wire to make into coil springs, titanium billets to make into jet engine parts, and many other situations. Some will be addressed at length later in the book. In fact, it is theorized in the antiques trade that many things such as wind-up toy trains and double-action cap pistols would have survived, had the spring material been tested for defects.

Another class of inspection installations that is necessary and cannot be faulted as fallacious can be characterized as follows:

1. Our process produces an invisible latent defect at random.
2. This defect would have dire consequences.
3. An electronic inspection method could detect this defect.
4. Therefore, install the inspection.

The key to this scenario is the concept of a process producing a nonconformance at random, not by a time-dependent degradation or a discernable root cause. Examples will be treated later in the book.

The argument for a third class of inspection installations runs like this:

1. A failure modes and effects analysis (FMEA) shows that certain detrimental occurrences may happen to production, yielding nonconforming parts.
2. We do not know when nonconforming parts will begin to be produced.
3. When they are produced, they may not be detected for a protracted time.
4. During this time much nonconforming material will be produced.
5. Entering into production downstream (or being sold), this nonconforming material will have undesirable consequences.
6. An automated 100% inspection method could detect and quarantine a very high percentage, for instance 99.78%, of this material without delay as-produced at a cost much lower than the consequences predicted by the analysis.
7. Install the inspection.

Great numbers of NDT installations have been made on the basis of arguments like the third case. In the quality assurance regime of the 1940s through the 1980s, it was necessary to install such equipment because there was no other viable way in use to ensure that the material going further into production would be good. That is not to say that other methods were not available. See, for instance, W. A. Shewhart (1931) and Western Electric Co. (1956) for statistical methods. The methods were not widely accepted or implemented even though they were known by certain people.

It remains to determine whether there is a fallacy in the third argument, and whether anything could be done to eliminate the logical error. Should the test have been installed?

2.9 The Root Problem: Out-of-Control Processes

Modern quality assurance, invented by Shewhart (1931), systematized by the Western Electric Co. (1956), and championed by Deming (1982), insists that the third argument in Section 2.8 is fallacious. The thesis is that nonconforming material is produced when and after a process goes out of control. The modern method addresses the points in Case 3 above as follows:

1. The FMEA results should be addressed by "continuous improvement" such that the process reaches high enough capability to produce only good material while under control.
2. We still do not know when it will go out of control, but Statistical Process Control "run rules" signal the failure relatively quickly. (These will be discussed in Chapter 3.)
3. The process is stopped; it does not continue to produce nonconforming material for a protracted time.
4. Only a moderate amount of unacceptable material is produced.
5. The material output from the time of detection back to the beginning of the "run rule" effecting the detection is set aside.
6. This material is inspected, salvaged, or junked.
7. Fix the process and continue production.

As one can see, modern Statistical Process Control depends upon detecting the onset of an out-of-control condition in a process rather than depending on mass inspection. In fact, one of Deming's Fourteen Points (Deming, 1982) to be explored in Chapter 4 is that "inspection is taboo." He noticed, as above, that management became addicted to inspection. He noted and decried the management's tendency in the early years to accept the argument in Case 3 in Section 2.8.

The next chapter deals with the operations of Statistical Process Control sufficiently to familiarize the reader with the subject. It does not go into the detail shown in books strictly on that subject, of which there are many, e.g., Shewhart (1931) and Western Electric Co. (1956).

Subsequent chapters give financial methods and examples showing that in some instances of great importance the dependence upon mass inspection can be proved to be viable, cost-effective, and profitable. The financial calculations are rigorous and can be repeated whenever the chance presents itself that Continuous Improvement may have made the inspection unnecessary.

3

Out of Control, Under Control, and Achieving Control for Processes

3.1 Out of Control as a Question of Information

In a factory, a process is the entity that acts upon raw material or upon an unfinished part to transform it to the next stage, nearer to becoming a completed part or a completed product. As such, a process has inputs and outputs. A process is a systematic set of actions involving men, machines, materials, and methods operating in an environment. All these factors may be thought of as inputs to the process. The process takes one of its inputs, generally a material, and does something to it to generate an output that has some value added to that input. It is intended that this one value-added output be a high-quality, useful entity, and that other outputs like metal chips, used fluids, pollution, and noise be containable. Generically, a process is represented in Figure 3.1. While a process may exist outside a factory, such as the shoveling and bricklaying analyzed by Frederick Winslow Taylor and recapitulated in Chapter 2, we are concerned chiefly with the process of doing manufacturing in factories.

Note that it is said that the process expressed in Figure 3.1 is "doing the manufacturing." The old definition of the "manufactory" in Chapter 2 is no longer operative. The *manus* part, signifying the human hand, is no longer critical to the making of things in a factory. The *process* makes the things. The Four Ms in Figure 3.1—men, machines, methods, and materials—are all in the process, but may be somewhat interchangeable. Even the environment may be adjusted. The methods are supplied by management as Taylorism required. Materials have always been involved. Machines may do more or less work than the men. Usually the men just watch the machines or perform minimal actions that are inconvenient to engineer into machine design as in Henry Ford's mass production. In the area of statistics and total quality management (TQM; see Chapter 4), all four of the Four Ms (and even the environment) are sources of the root causes of errors in the processes. The change of man from a talented and irreplaceable master to a detrimental source of errors was made by Taylor and Ford and is essentially complete.

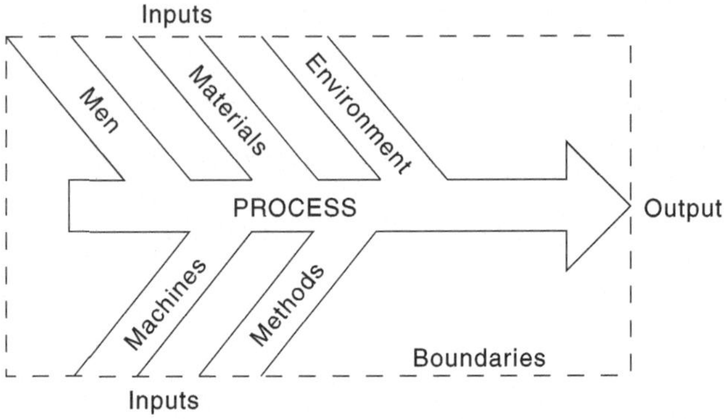

FIGURE 3.1
Principal vertebrae of a process fishbone chart defining the possible variables: men, materials, machines, methods, and environment. The process has an output and may go out of control because of perturbations in the five variables.

Within a factory at any point in time, a process is under control or out of control. It is vital to understand the concept of being *under control.* Process control is often thought of as adjusting inputs according to some read-out mechanisms so that the inputs, such as voltage and fluid flow, are as specified by the process instructions. However, this is not enough. The voltmeter may drift, that is, go out of control, so that the controlling mechanism becomes incorrect and the process goes out of control. If a process goes out of control, the quality of its production degrades. Some final arbiter must be provided to prove that the process was actually under control from time A to time B. The purpose of this chapter is to provide and explain one empirical/mathematical final arbiter. The critical skill is no longer an expert man but rather has become mathematics—a method. Man is at the bottom of the heap in the Four Ms.

The diagram of a process in Figure 3.1 is perfectly general. One may suppose that the process was designed by certain men, typically industrial engineers, who chose a factory environment and decided upon certain methods that would be embodied in a machine that other men would have to operate or at least watch over for a period of time, consuming some materials and operating constructively on one type of input material, making something we shall call a *part*. Let us suppose further that the industrial engineers operated the new process for a period of time using good materials, and ascertained that all the parts turned out by the process were acceptable. Then, after writing up work instructions, they turned the process over to the line supervisor to staff and run. How is this process analyzed by Taylor, by mass production exponents, and by more modern quality managers?

In the Taylor milieu, this process should produce good parts forever while needing only some maintenance on the machine. This assumption was also made regularly in factory work by the mass production philosophers. In reality, what happens?

The reality is that the process will go out of control at some unknown time in the future and begin producing unacceptable parts. Going out of control is itself a process and must be guarded against. One does not know, *a priori*, when the process will go out of control or what the nature of the failure will be. The mathematical final arbiter of in-control versus out-of-control must be independent of the individual method of going out of control known as the *root cause*. The arbiter to be discussed in this chapter is independent in just this necessary sense. What, then, is the nature of *going out of control*?

The perturbation disturbing control is generally statistical because all the inputs in Figure 3.1 are prone to statistical fluctuations. Blame is not a proper approach to attacking an out-of-control condition. If perturbations to processes happen at random (statistically) like tsunamis, then one cannot blame a person for the fact that the process went out of control any more than one may blame a person or God for the multiple deaths in a flood. The man in Figure 3.1, no longer a good factor, is not a bad factor either. Management is to blame for not having detection means installed, of course. The means of detection for the process are information and statistics implemented in a certain systematic order.

The first requirement is information. When did the process go out of control? When did it start to go out of control? How do we get this information? As this section is entitled, being out of control is a question of information. One does not want to wait two weeks until 50,000 faulty parts have been produced to take some corrective action. When does out-of-control begin, and how does one detect it?

3.2 Statistical Process Control (SPC) To Get Information

The mathematical/empirical arbiter of in control versus out of control is *statistical process control* (SPC). The modern emphasis is to use SPC to keep processes under control. However, keeping processes under control is a fallacy. Can SPC keep processes under control? No. Nothing can *keep* processes under control. Processes inevitably go out of control. When a process begins to go out of control, it begins to produce nonconforming parts. After a process is out of control, SPC can tell you that it *has* gone out of control. This is in the past tense—after the fact. But how long after the fact? That depends on the frequency at which samples are taken for the SPC calculations. Is the period every hour, every 4 hours, every shift? Besides, SPC is statistical itself. It can tell you, for instance, that it is only 1% probable that your process is still under control.

Going out of control is itself a process. The process of going out of control may be gradual in a sense that will require several of these chosen periods before the SPC test will signal the out-of-control condition. You may have to wait 5 or 8 of the 4-hour periods, for instance, to be 99% sure that the process is out of control. That is, after the process begins to go out of control, it may require 5 or 8 of the time periods before you have only a 1% chance, according to the SPC control charts, of still being in control. Only then will you be willing to stop the process and repair it. That is the key to using SPC: wait until it tells you the process is probably out of control; then stop it and fix it. This is the SPC function of getting information.

Having stopped the process, you must quarantine the production back to the beginning of the gradual process that has taken it out of control. The parts made during this period of time must be tested to ascertain that they are good or that they should be reworked or scrapped. Some will be good; some must be attended to. The batch cannot be shipped without testing. This is a limitation when using a just-in-time inventory. If your process is making several hundred parts per hour, then a much larger batch of material cannot be shipped, and all of these parts must be tested.

All of the SPC processes and procedures alluded to here are completely explained in W. A. Shewhart (1931) and Western Electric Co. (1956). These texts should be studied in depth to understand the use of SPC. A few more necessary details will be given below to make SPC more intelligible.

To reiterate, SPC does *not* keep a process under control. A process will inevitably go out of control. SPC is needed to tell you to a degree of certainty (such as 99%) that the process is finally out of control.

3.3 A Review of Statistical Process Control

SPC, still in use today, was derived and developed 15 years before the explosive growth of modern electronics for civilian industrial purposes, which can be dated between 1942 and 1946. The assumptions of SPC include this: **Measurements will be made by hand by laborers who will measure extrinsic physical properties of manufactured objects.** A laborer might measure the diameter of five shafts using a micrometer or the weight of five bags of sugar using a scale. An intrinsic measurement like tensile strength or sweetness was not accessible then. It may be procurable today with electronics, but not then (Shewhart, 1931). The reintroduction of SPC by W. E. Deming (1982) was based on the same scenario—laborers would measure extrinsic properties of parts manually to do SPC on the parts-making process.

Five is not a magic number, but is a typical number of parts to be measured in each time period. One would measure the last five parts made in that time period. This number, which may be chosen for convenience, is generally denoted as n. The time period is typically one hour, 4 hours, or one shift. Typically, $n = 5$ successive parts, which are measured at the end of each time period. Some variable X is measured. No individual one of these values of X_i is used to signal an out-of-control condition, but rather two statistics calculated from the measurements are used in an algorithm. The two statistics are typically the Mean X-Bar and the Range R (maximum minus minimum values). Other statistics are possible, such as proportion defective, but these are left to the student to find in the textbooks as needed. In equation form, X-Bar and R are as follows:

$$\text{X-Bar} = \frac{X_1 + X_2 + X_3 + \ldots\ldots + X_n}{n} \tag{3.1}$$

and

$$R = X_{max} - X_{min} \quad \text{(among the n specimens)} \tag{3.2}$$

So what do we do with these statistics?

The mean and the range are to be compared with *control limits* to determine whether the process has gone out of control. These control limits are drawn on *control charts* on which the values X-Bar and R are plotted at each subsequent time period. For each statistic, there will be an *upper control limit* (UCL) and a *lower control limit* (LCL). The statistics must stay within the control limits to a very specific degree to indicate a process under control. In particular, the control limits on the mean are *not* the upper and lower specification limits on the part. The control limits are much tighter than the specification limits. The control limits are calculated from the grand mean, X-Double-Bar, of the means of many sets of n samples and from the mean of the ranges, R-Bar, of the same group of many sets. Many sets could typically be twenty or more, but never less than ten (see Western Electric Co., 1956).

The control limits have been derived mathematically. They depend upon the values of X-Double-Bar and of R-Bar. Multiplying factors for the calculation of the control limits have been derived from theory and are shown, for instance, in Western Electric Co. (1956), on page 12. The multiplying factors are functions of the number of observations n in a sample. In this chapter, just the set for the useful case $n = 5$ will be used. These are

$$A_2 = 0.58$$

$$D_3 = 0.00$$

$$D_4 = 2.11$$

How are these multipliers used to find the control limits on the control charts?

The value of X is measured for each specimen in the large number of groups of n specimens. Twenty groups would be typical. Then R is calculated

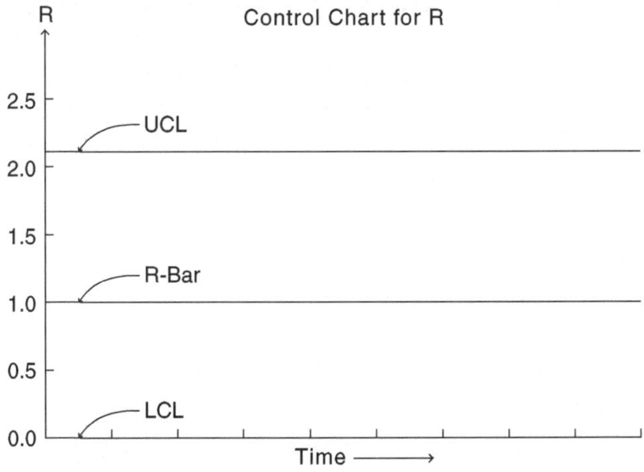

FIGURE 3.2
Control chart for range (R) with mean and upper and lower control limits.

for each of these groups. After the last group is processed, the average R-Bar is calculated. The two control limits on R are given by

$$\text{LCL(R)} = D_3 \times \text{R-Bar} \tag{3.3}$$

and

$$\text{UCL(R)} = D_4 \times \text{R-Bar} \tag{3.4}$$

The two control limits and the (asymmetric) centerline R-Bar are drawn on a graph with time as the abscissa. (See Figure 3.2). This graph is drawn with R-Bar = 1.0 and $n = 5$ with the multipliers above to fix ideas. To effect the actual control of the process, the values of R will be plotted on this graph as production goes on, and more sets of n specimens are measured after each time period. A control chart is also needed for X-Bar. It is calculated as follows.

After the R-chart is set up, an X-Bar control chart must be set up. Its centerline will be the value of X-Double-Bar, the average of all the X-Bars from the large number of sets. The two control limits on X-Bar are given by

$$\text{LCL(X-Bar)} = \text{X-Double-Bar} - [(\text{A2}) \times (\text{R-Bar})] \tag{3.5}$$

and

$$\text{UCL(X-Bar)} = \text{X-Double-Bar} + [(\text{A2}) \times (\text{R-Bar})] \tag{3.6}$$

These two control limits are drawn on another graph with time as the abscissa, (also see Figure 3.3). To fix ideas, we use the same R-Bar of 1.0 and the same n of 5 as in the R-chart. In Figure 3.3, X-Double-Bar is taken as 10.0. (One can see that this choice is an exaggeration because a 10-pound bag of

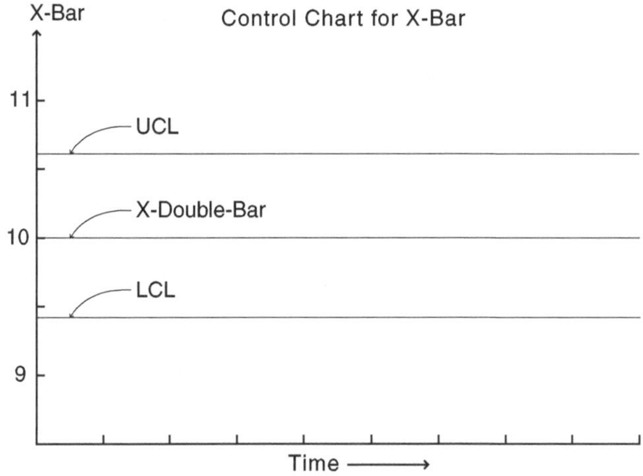

FIGURE 3.3
Control chart for X-Bar with mean and upper and lower control limits.

sugar ought to be filled more accurately than the range between 9.5 and 10.5 pounds.)

The industrial engineers mentioned in Section 3.1 made the error of turning over the process to the line personnel before carrying out all the above operations to generate control charts. In addition, the engineers should have provided information on the ways the laborers should interpret the activity of the points being entered onto the control charts over time. The modern expectation is that the laborers would do the simple measurements and arithmetic every 4 hours, for example, enter the two resultant points onto the two graphs, and be trained to recognize unusual meandering of the points over time. In the modern factory they would be empowered to stop production if the meandering of the points indicated an out-of-control condition. At least Deming under TQM (Chapter 4) intended to empower them. Thus, the laborers would psychologically regain at least part of their control over their work environment and output, which had been taken away by Taylor and Ford. For one thing, they would be carrying out another set of work instructions in addition to the work instructions that control their production work within the process. Of course, Taylor could have written such instructions if he had known statistics. Empowerment to stop production would be a positive feeling not offered by Taylor or Ford.

We have mentioned the meandering of the statistical data points. What do they do, quantitatively?

In general, the statistics X-Bar and *R* will fluctuate around the middle lines of the charts. Moderate fluctuation in a random fashion is to be expected and does not indicate an out-of-control condition until certain conditions or trends become apparent. The simplest situation indicating an out-of-control condition is for X-Bar or *R* to fall outside the control limits. One instance of

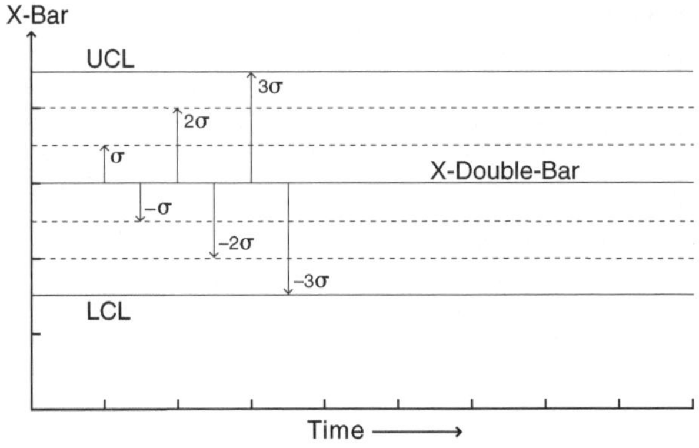

FIGURE 3.4
Control chart for X-Bar with control limits divided into six bands for run rules.

exceedance indicates that there is less than a 1% probability that the process is still under control. In reality, the multipliers listed above were derived to give just such a result. The width from the middle line to each control limit is essentially three standard deviations of the process. Only 0.13% of a bell curve lies in each tail beyond three standard deviations from the mean, so it is highly probable that an excursion into that fringe of the tail would be abnormal. Are there other abnormal conditions? Yes.

If one were to divide the area from the centerline to the control limits into three equal bands, each would be about one standard deviation sigma (σ). An X-Bar chart divided into six bands like this is shown in Figure 3.4. Other rules can be derived involving many successive points being outside one or two standard deviations, that is, falling into these bands. The rules also show conditions in which the probability that the process is still under control is less than 1%. These rules are termed *run rules,* which means that as the process is running along, a sequence of statistical points run up or down in a particular fashion, which can be formulated as a rule. The four run rules advocated by Western Electric are given in Table 3.1. These are called Test 1 through Test 4 for instability (Western Electric Co., 1956, 25–27). Other run

TABLE 3.1

Western Electric Run Rules for Out-of-Control Conditions

1. A single point outside three sigma (3σ).
2. Two out of three successive points outside two sigma (2σ) on one side of the centerline.
3. Four out of five successive points outside one sigma (1σ) on one side of the centerline.
4. Eight successive points on one side of the centerline.

Source: Western Electric Co. (1956). *Statistical Quality Control Handbook.* Western Electric Co., Newark, NJ, pp. 25–27.

rules are possible. The Ford Motor Company, for instance, advocated another set after adopting the Deming management method around 1981.

It is not known that the process is out of control until the end of the run rule that detects the out-of-control condition, but the logic of the run rule indicates that the process was actually out of control during the production of the entire set of points used by the particular run rule to make the out-of-control call. Using the first rule, the time for one point was expended. Using the second rule, the time for two or three points was expended. Using the third rule, the time for four or five points was expended. The fourth rule expended eight time slots. All the production made during those expended periods of time must be considered to be out of control. How does the machine operator find these points and make a decision about an under-control or out-of-control condition?

The machine operator should make the requisite measurements and calculations as time goes on, and faithfully plot the points on the control charts immediately. His alert observation of the behavior of the points as interpreted by the run rules, which he keeps at hand written down or has memorized, will tell him when the process has gone out of control. Then he should have the authority to stop the process and undertake corrective action. Corrective action includes quarantining the parts made during the run rule detecting the condition. Note the definitions of *corrective action* discussed later in Chapter 5 on International Standardization Organization (ISO)-9000.

Our industrial engineers should not have considered their job complete until the line operator felt comfortable with the control process above. The line operators can now use their intelligence and willpower in maintaining quality of output. Part of the outlook of the journeyman is reinstated toward pre-Taylor times.

3.4 Automated Run Rules with Computers

Since 1987 it has been possible to purchase automated equipment to perform SPC run rule analysis with automated nondestructive testing (NDT) measurements. Systems that operate under computer control are available to do two functions simultaneously (K. J. Law Engineers, Inc., 1987; Perceptron, Inc., 1988). First, the computers control the NDT equipment and command the data acquisition. Second, the computers, using run rule algorithms, pick points from the data stream and compute the occurrence of an out-of-control condition, flagging it. Some other computer programs are available that can be interfaced with inspection equipment on a custom basis (Advanced Systems and Designs, Inc., 1985; BBN [Bolt, Beranek, and Newman] Software Products, Inc., 1986). E. P. Papadakis (1990) has written and reported on a program that can automatically perform the Western Electric run rules (see Table 3.1). The author also attached a program to simulate a process's going out of control to demonstrate how rapidly the automated run rules could detect out-of-control situations.

The run rule program operates on the data simulation to do many calculations including, of course, statistics. It was confirmed for the benefit of management that an automated run rule program could effectually do automated SPC.

It remained to be determined how factory workers could interact with these programs and systems in order to feel empowered and intelligent.

3.5 Statistical Process Control Results as Statistics

It is pretty obvious that SPC results are statistical in themselves. As the results of manufacturing may fluctuate, the results of the SPC used upon manufacturing may fluctuate. Unknowns intrude. It may be that the results from five successive parts coming down the line may differ from the results on the next five. A resultant X-Bar might be away from the centerline by 2.9 instead of 3.1. Some results may retard or accelerate the apparent detection of an out-of-control condition. This will not have a great effect in the long run, but should be considered as one tries to use SPC in an absolute sense. Continue to remember that statistical process control is statistical. Perhaps an input, unchecked, has untoward effects upon an output. Deming (1982) devotes a long chapter (see Chapter 13) to the possible need for testing incoming material to eliminate fluctuations in output. The ideas can be better understood through examination of Figure 3.1.

The original fishbone diagram, Figure 3.1, can lead to analyses of things that might go wrong. Each input arm can itself have multiple branches, each potentially producing a problem. Many interesting unexpected perturbations to processes have been uncovered by brainstorming sessions and astute analyses. In one case, a black contaminant crept into a white yarn vat every noon beginning in June of one year (Papadakis, 1974). It was discovered that the crew of a diesel switch engine had begun parking it in the shade of the back wall of the mill to eat lunch. The air intake for the yarn machine was just above the diesel exhaust. In another case, a high-tech machine was installed in a factory with skylights. On sunny days the thermal expansion of the bed of the machine was great enough to put its production out of control. Environment as a statistical input can be very fickle. Fickle is just another word for statistical. Is there any systematic way to attack anomalies like these and find the root causes expeditiously?

When unknown extraneous causes like these come up, a control chart can be used as an analysis tool to permit engineers to discover root causes of problems because of the systematic types of errors that show up. Teaching this analysis is beyond the scope of this book. A very complete text on the subject is provided in Western Electric Co. (1956). The student should be aware of the possibilities. Much SPC effort is directed toward problem solving as well as problem detection. Often the distribution of observed X-Bars, other than Gaussian (bell curve), yields clues as to the causes of the differences from ordinary statistics.

3.6 Out-of-Control Quarantining vs. Just-in-Time Inventory

When you find an indication that a process has gone out of control, what should you do? Quarantining is the answer. The parts should be put in the "sick bay" and inspected—analogous to taking their temperature.

The process of using the run rules to detect out-of-control conditions was explained earlier to mean that the process was actually out of control throughout the operative run rule. The length of time could be as long as five to eight periods between sampling tests. Each period could be as long as one shift or whatever time had been chosen by the responsible engineer. That means that the company should be prepared to quarantine all the parts made during the most recent eight time periods (the fourth run rule). Extra parts should be ready for shipment to cover orders represented by the eight time periods plus the probable time for repair of the process. That would guarantee just-in-time inventory shipments at the output of the process. Note the possibility of a time delay if you chose to operate without the extra inventory. It might be called just-in-case inventory, but it is necessary.

What is the scenario after detection of an out-of-control condition? Repair and restart. If the time for repairs were to be one or two day shifts, with the night shift also covered because the engineers would not be there to do the fixing, so be it. Just-in-time inventory shipments presuppose a continuous flow of acceptable parts off an under-control process, so the shipments must continue. "The show must go on," as they say in the circus. For the shipments to continue, they must come from the extra parts ready for shipment mentioned above. This cache of parts is what the statisticians disparagingly call just-in-case inventory. Certain companies have been convinced to do away with just-in-case inventory. Upon process failures, they have been found lacking or caught napping. One company has been known to fly parts cross-country by Flying Tiger Airlines at great expense to meet production schedules rather than expend the interest on the money to pay for just-in-case inventory and its storage. This was thought to be cost effective during double-digit inflation.

If one recognizes that his company needs just-in-case inventory to accomplish just-in-time shipments, then production can proceed smoothly. One should also recognize that SPC is the mathematical/empirical arbiter of conditions of in-control vs. out-of-control conditions. It will be shown in Chapter 6 that SPC should be used as a preliminary screening process before financial calculations are made about installing 100% inspection with high-tech methods.

4

Total Quality Management with Statistical Process Control and Inspection

4.1 Total Quality Management and Deming's Fourteen Points

Total quality management (TQM) is a complete and self-contained system of management based on the lifetime philosophy of Dr. W. E. Deming. It is Deming who characterized it as complete and self-contained, and his disciples think of it as such. It is certainly a philosophy of management, and it certainly contains many facets not found in the management styles and schools of other quality professionals. It contradicts some of the tenets of Frederick Winslow Taylor and Henry Ford, and it was a major *coup de grace* for Deming to have his philosophy adopted by the Ford Motor Company in 1980. While at Ford, the author studied under Dr. Deming and under his chief appointee for corporate quality.

Statistics were keys to the progress of the philosophy, as it had been to Deming's career since his use of statistics in the 1930 United States census. (Congress is further behind than 1930, unwilling to use statistics to count the homeless to this day. The question comes up every decade when the political party, willing to assist homeless and helpless people, seeks favorable redistricting for congressional seats.) Detection of out-of-control by statistics is at the core of Deming's thought process about quality.

How did Deming win acceptance in the United States, given the predominance of the manufacturing philosophies of Taylor and Ford?

Deming's regime of statistical process control (SPC) following W.A. Shewhart (1931) was accepted by Japan during its rebuilding after 1946. Deming was the major consultant for Japan on industrial quality. His work turned the image of Japan as the maker of junky tin toys into the manufacturer of superlative automobiles. Indeed, Japan initiated and issued the Deming Medal for quality accomplishments in its own industries. Japan invented some techniques such as "quality circles," which countermanded the Taylor philosophy of "kicking all knowledge upstairs." In quality circles, some of the knowledge and thinking power reside with the laborers. They identify quality issues, isolate root causes, and solve the problems. In the mean time, the United States was dismantling the SPC effort that had been available since 1930 (Shewhart, 1931) and exquisitely expressed in 1956 by Western Electric, where it was still being used internally. The United States was going back to the old Taylor and Ford deterministic production ideas. Statistics was the golden key enabling excellence to be tossed into a dark lagoon. And, as it turned out, Japan was the creature that arose from that dark lagoon—the remnants of its Greater Asia-Pacific Co-Prosperity Sphere. By the 1970s the Japanese had taken over the manufacture of all zippers, diaper pins, transistor radios, essentially all television sets, and other home electronics. By the end of the 1970s, Japan had made inroads into the auto industry, equaling 30% of American production. Japan was "eating our lunch," as it was termed in the automobile industry.

When it was realized by some in the United States that the Deming methods had given Japan an advantage in manufacturing, American industry belatedly began to seek out Deming for direction. Deming portrayed himself as the savior of American industry. As the sole source of his own successful philosophy, W. E. Deming cut enviable deals with these industries. He required a commitment from a company sight unseen, to adapt his philosophy and methods completely and unquestioningly before he would give even the first lecture. This is similar to Taylor, who required that "the organizer be in a position of absolute authority" (Parkhurst, 1917, 4). Deming required a commitment on the part of the company to teach all its personnel his methods and to teach its suppliers, too. At a large multinational firm, this meant having Deming himself teach four-day courses to thousands of employees for several years. (I took the course twice.) As follow-up, people of Deming's choosing were installed in positions in new quality organizations within the multinational to keep the work going.

To Deming, the philosophy seemed self-contained and complete. Gaping holes were visible to many attendees. This book fills one of those holes. However, it is important to understand the Deming approach just as it is important to comprehend the Taylor method of scientific management and Ford's mass production.

The Deming approach is embodied in his Fourteen Points patterned after Woodrow Wilson's 14 Points. The Fourteen Points form essentially a table of contents to Deming's mind. These Fourteen Points should be studied directly from the source so that the student will understand the exact denotation and connotation of the phraseology (see Deming, 1982, 16, 17–50). An exposition of the Fourteen Points with succinct explanations is given in M. Walton (1986b), pages 34 through 36.

I have attempted to distill the main idea of each of the Fourteen Points into a key word, or at most, two key words. Given the key word as the beginning of a thought, one can expand it into a family of thoughts and examples encompassing the meaning of the point with respect to modern industry. In fact, some of the diction in Dr. Deming's original formulation is somewhat delimiting (limits one might wish to escape). For instance, the phrase "training on the job" is used in Point 6. It happens that on-the-job

TABLE 4.1

Key Words for Deming's Fourteen Points

1. Decision: Improvement
2. Decision: Enforcement
3. Inspection: Taboo
4. Suppliers: Good-not-Cheap
5. Improvements: Pinpointing
6. Training: Modern
7. Supervision: Modern
8. Fear: Taboo
9. Teams, Not Barriers
10. Slogans: Counterproductive
11. Quotas: Taboo
12. Workmanship: Pride
13. Education and Training
14. Implementation: Staffing

training (OJT) means something much different in the airline maintenance inspection industry than it means in some other venues. My key words in this case are Training: Modern. The key words, as jumping-off points for the interpretation of Deming's Fourteen Points, are listed in Table 4.1.

Inspection shows that many of these key words deal with human resources and interpersonal relations. Expanding the meaning of each one, however, there is an insistence upon the relevance of statistics, and in particular, SPC. The concept of inspection is treated in the meaningful interpretation of several of these points. Modern ideas of inspection will be interspersed to augment Deming's fundamental statistical opinions. These points will be summarized briefly as they are important to the subject of this book, the financial justification of nondestructive testing (NDT). NDT is a family of relatively modern methods for high-tech inspection. As will be seen as we proceed, Deming had some opinions that clash with NDT. In addition, he held opinions formulated before the development of many inspection methods. These new methods may actually supersede the detrimental aspects he saw in old-fashioned inspection. The relevant parts of Deming's Fourteen Points will be explained thoroughly.

4.2 Deming's Fourteen Points Taken Sequentially

4.2.1 Point 1 Key Words: Decision: Improvement

The company planning to adopt Deming's methods had to sign on by making a fundamental *decision* to be faithful to the Deming philosophy for the long haul before Deming would sign on to accept consulting fees from them and to teach them. The chief executive officer (CEO) and the board of directors had to agree to be faithful. The decision had to be adopted as a "religious conversion" of the secular company. The decision was that the company was committed to improving its quality and way of doing business. The implication was that this was irreversible. Deming insisted that his methods were more important than the bottom line each quarter. To have a going concern, he said, it was necessary to have this unswerving purpose from year to year so that 3 or 10 years out, the company would still be in business while its competitors, who had worried about quarterly profits, would have foundered.

The main improvement had to be in quality. The title of Deming's principal book, *Quality, Productivity, and Competitive Position* (1982), is to be interpreted as follows: If you *raise quality,* then *productivity will increase* because of less waste (rework); productivity increases, along with an improved quality image (reputation) will *raise revenues* (more sales), which can be spent on whatever is needed to make your competitive position stronger vis-à-vis the other companies in the field.

This point has several corollaries or subsidiary explanations as follows:

- Industry must admit to itself that the Taylor scientific management method and the Ford moving production line overlaid the potential efficiency of the production line with poor quality, bad work ethics, inefficiency, and high costs.
- This detrimental overlay cannot be overcome overnight with a sugar-coated pill; the crisis can only be solved by long-term resolve.
- This resolution to do something about the problem requires unswerving direction with this purpose in mind.
- The requirement is an improvement in quality of both products and services. This improves both image (external view of quality) and productivity (internal quality with less rework waste).
- The improvement must be carried on constantly and purposefully because faltering causes backsliding and the competition is continuously improving. Comparisons with industry competitors come annually from J. D. Powers reports and so forth.
- The main purpose is to stay in business by remaining competitive. Plan ahead. Not just profit this quarter.
- Improvement is a *war aim* because trade is war carried on by other means (to paraphrase Bismarck), and trade is international competition.
- With all competitors holding an unswerving determination to improve and applying the correct methods, all competitors will strive, asymptotically, toward the same high quality level approaching (but never reaching) perfection, and competition will be on a level playing field with respect to quality. Those who do not improve quality will fail.

This unswerving determination must be started somewhere, so it is necessary to adopt the new philosophy as covered in the next point.

4.2.2 Point 2 Key Words: Decision: Enforcement

Once the CEO has made the decision to be faithful to the Deming philosophy, it is his job to enforce his decision upon the entire leadership of the company from the chief operating officer (COO) on down. All must be faithful, and all must be trained.

- This is a new philosophy; it must be adopted (a) as a whole, like a religion (Deming's own words, 1982, p. 19), (b) not piecemeal, and (c) accepted by everyone in the company.
- The chairman of the board and the COO must become convinced and must bring all executives into compliance. Note the parallel in Taylor's "absolute authority" of the organizer; otherwise, Deming refused to work for the company.
- All the personnel in the company must be educated in the philosophy and forced to apply it.
- All the company's suppliers must be forced to work under the philosophy, at least on product to be supplied to the company, or else be dropped from the bidders list.
- While everyone is forced to work under the philosophy, everyone is actually expected to adopt it internally.
- To summarize, the good New Year's resolution to create unswerving purpose is no good unless you adopt it, implement it, and carry it out.

Next, Deming moves into remedial action for a supposed flaw in the old procedures.

4.2.3 Point 3 Key Words: Inspection: Taboo

Inspection to ensure or produce high quality is considered taboo in all but the most limited circumstances. Inspection is to be eliminated except in a handful of situations. Deming noted that companies had developed a dependence upon inspection to make sure only good material was shipped out the door, but that the companies had neglected many of the steps that could actually produce good quality. Deming thought of this dependence on inspection as an addiction or as analogous to a codependent personality. Some of the inspection scenarios leading to the inadequate addressing of poor quality were given in Chapter 2.

Dependence is the key word here. Companies had become dependent upon inspection when they realized that they could not produce quality consistently, but were required to ship quality output. Mass inspection, Deming's principal taboo, means inspecting essentially everything. Mass inspection, as explained in Chapter 2, was not tied to any feedback to the process or to the design. Periodic measurement through SPC, on the other hand, was thought to be good because its purpose was feedback and because statistics was Deming's specialty.

Point 3 of the Deming philosophy has wreaked havoc with the NDT industry because statisticians at companies have taken this injunction literally, at Deming's behest. They have acted to destroy inspection without doing the rigorous financial calculations taught in this book. Deming's opinions with respect to dependence upon mass inspection are based upon years of observing messy management practices:

- Reliance or dependence on mass inspection is the demon in Deming's pantheon of evil mind-sets of management.
- Interpretation: Having mass inspection means you plan to make errors. You plan to make garbage and catch it later. Deming believed inspection encouraged carelessness.
- Relying on mass inspection means that you are not trying hard enough to do it right the first time.
- On the other hand, Deming was the first to admit that it is statistically impossible to achieve zero defects. All processes and human activities are statistical. Sometimes outliers will happen and occasionally (inevitably) processes will go out of control. See Figure 3.1 and its explanation.

Deming condemned the behavior of management in employing inspection personnel—planning to make garbage, intending to make errors, being paid to be deliberately careless, and not trying hard enough to do it right while sweating bullets to pull the company out of a bind. (The manager and the inspector could hardly have thought well of a guru who charged them thus, and some technologists in client companies became hostile.)

And yet this is precisely what Deming perceived when he looked at a company on a consulting basis. He saw inspection means, whether manual, visual, or electronic, applied to the outputs of processes without effort being expended to ensure that the processes were under control. He saw highly efficient inspectors throwing away parts without the feedback to the operator that nonconforming parts were being made. He saw information garnered on outputs wasted because the inputs were at fault, not the process. He saw that the company was interested in shipping good parts but was not determined to make only good parts (or at least the best possible parts considered statistically). He saw that the company was not viewing processes statistically. The company did not see the process as yearning to be improved. Deming saw that a company would be happy to spend three weeks of an engineer's time on research and development (R&D), $3,000 on an electronic box, and half a man-year of labor annually to ensure that no bad copies of a certain part were shipped, rather than determine and fix the root cause of the poor quality of that part. Possibly a new, expensive furnace was needed; its purchase might have solved many problems, but reliance on mass inspection was easier to justify with the management outlook at the time. Perhaps training was needed, or perhaps a new rule for cigarette breaks. No one discovered the root cause, but inspection was adopted. The inspection engineer could not determine the problem because of the barriers between staff areas (see Point 9). Also refer back to Taylor's deliberate planning of barriers covered in Chapter 2.

If the reader thinks that the foregoing analysis including the example is imaginary, it is not. I was the supervisor of the group ordered to develop the test in question with the $3,000 electronic box. Success was considered valuable to the company, as indeed it was, given the milieu of the moment. The integrity of the heat treatment of the parts in question had to be ensured and bad parts rejected because otherwise the parts could break and cause parked automobiles to roll away, causing accidents. This test is one of a multitude treated similarly by management. Interestingly, the test was for an intrinsic variable yielding a latent flaw that could not have been found by statistical measurements on extrinsic variables.

To his credit, Deming acknowledged that inspection should be done at least for a time in certain circumstances. These situations are as follows:

- If safety is involved. See Delta Items specified for automobiles by the National Highway Traffic Safety Administration. In these cases, inspection should be continuous, 100%, and forever.
- If a process is new or changed so that statistics must be gathered, Deming (1982) suggested testing for six months.
- If a process makes parts, each of which is unique, so that the process cannot be considered under control. See the description of instant nodular iron in Chapter 9, for instance.
- If inspection is cost-effective even when a process is under control.

The last circumstance is basically the topic of this entire book. It turns out that W. E. Deming mentioned this idea in his lectures and wrote it up in the before-publication notes (1981) for his book (1982). By the time the book was published, the derivation of the proof for the idea had been relegated to a problem for the student (Chapter 13 of Deming's book includes a discussion of this issue). As the principal body of students was composed of busy engineers, the idea slipped through the cracks. I published an understandable derivation, an explanation of its implementation, and several industrial examples in a paper (Papadakis, 1985) in a journal in the quality control field. The statisticians employed by companies under the direction of Deming generally neglected the topic, preferring statistics.

Deming downplayed the need for inspection and downgraded the image of inspection in order to accelerate the implementation of his dictum about ceasing reliance on mass inspection. While it was a good thing to get processes under control, the idea of inspection while under control was deemphasized in his lectures and by his chosen disciples installed in company positions. One striking omission in his philosophy concerned the detection of what he called "latent flaws." These are manufacturing errors that would be characterized by a physicist as intrinsic variables. They cannot be found by extrinsic measurements like diameter, or fitting a gauge, or weight, which could be measured by the human inspectors along a production line. "Latent flaws" such as excess hardness or inadequate tensile strength or internal cracks in extrusions can often be detected (after proof by research) by electronic means. Deming was ignorant of the possibility of detecting latent defects electronically. I questioned him on this at a meeting of his handpicked Statistical Methods Council at the Ford Motor Company (Deming, 1984). He said explicitly that he did not know about electronic detection of intrinsic variables. This means that he was working on assumptions made circa 1930, which would have let inspectors see only the tip of the iceberg among manufacturing errors. A company could have been sunk by the need to inspect for latent flaws, which he did not understand as tractable.

The position taken in this book is that inspection should be considered in every situation. Its advisability can be calculated mathematically using financial data. The question "Should we inspect?" can be answered rigorously, and may be yes or no (see Chapters 7 and 9).

Another piece of remedial action comes next.

4.2.4 Point 4 Key Words: Suppliers: Good, Not Cheap

The injunction is to find and settle upon good suppliers who can be trusted. They may not be the cheapest in the bidding war, but they will help your production in the long run. This injunction is contrary to the ordinary way of doing business. The new way should include the following:

- Looking for the lowest bidder is obsolete. "Price has no meaning without a measure of the quality being produced" (Deming, 1982, 23).
- One must look for the supplier who can deliver quality continuously. (Same as International Standardization Organization [ISO]-9000 emphasis.)
- One wants just-in-time (JIT) delivery of quality goods to go directly into your production line. (Henry Ford insisted on this way back in 1914 to feed his chassis lines. Of course, he was building all the subassemblies, so he could insist upon it).
- Make long-term arrangements with adequate quality suppliers. Save both them and you the hassle and uncertainty of bidding at every whim.

- Reduce the number of suppliers you deal with.
- Make the supplier responsible for supplying quality by qualifying the supplier (vendor) and relying upon a good relationship. Require the vendor to be responsible for quality.
- The vendor must be able to prove his quality by records and statistics such as control charts.
- The vendor should obey the Fourteen Points.
- Change the job of the buyer from seeking the lowest bidder to finding quality suppliers.
- Remember that the lowest price brings with it poor quality and high cost. "He that has a rule to give his business to the lowest bidder deserves to get rooked" (Deming, 1982, 23).

A most interesting concatenation of Points 3 and 4 occurred on my watch running the nondestructive testing group at the Ford Motor Company Manufacturing Development Center. A contract had been given to the lowest bidder by a major division of the company. The supplier was shipping faulty parts to Ford and was covering up its mistakes by a ruse that made inoperative the only visual and tactile method of detecting the faults. Quality could not be proven bad or improved without my group's first inventing an electronic inspection method for the (deliberately) hidden latent flaw. The following is the text of a short report written by the author (Papadakis, 2000b, 1031–1034) about this detection by inspection that saved more than $1 billion, which could have been the detrimental cost in the worst-case scenario.

Most of you as kids have glued plastic models together such as jet planes, Old Ironsides, the Nautilus, and so on. Full-size trucks are not much different, at least some parts of certain models. Major truck body parts like whole hoods with integral fenders may be molded in two or three sections and adhesively bonded together.

I ran into a problem with the bonds which held heavy truck hoods together. The right and left halves of these heavy truck hoods with integral fenders were molded of sheet molding compound (SMC) which is a thermosetting plastic resin containing about 30% by volume of chopped glass fibers (2 inches long) randomly oriented for reinforcement. The raw material comes in soft, pliable sheets which are cut to size, laid into molds, compressed to shape and thickness, and heated to cure into rigid complex shapes. These shapes, such as the right and left halves of a truck from the bumper to the windshield, are then bonded together with a thermosetting adhesive. The lap joint is typically at least 1 inch wide. The adhesive is supposed to spread throughout the joint area when the two parts are brought together and then is supposed to cure, holding the parts together. The parts in question were made by a first-tier supplier and shipped to a truck assembly plant for final assembly into vehicles.

Failures of the adhesive bond can occur from several causes, including (1) unclean surfaces, (2) lack of adhesive, (3) pre-cure of the adhesive if the parts are not put together soon enough, and (4) spring-back of the parts if they are not clamped into position during the cure. The problem I ran into was compounded by all of these causes, not just one. Contamination could never be ruled out because of the shipping and handling routine. Adhesive was applied by hand with things like caulking guns so that areas could be missed in a hurry-up routine. Workers could take a cigarette break between the application of the adhesive and the joining of the parts. Because the parts were not clamped but simply set aside, gravity and mismatch could cause parting of the adhesive line in the adhesive during curing at room temperature. And, compounding the problem still further, a relatively rapidly polymerizing adhesive was used so that the parts would not have much time to sag apart before curing. This attempt to circumvent the spring-back problem (without the use of clamping jigs) exacerbated the pre-cure problem if there were assembly delays.

The problem showed itself in the field where fleets of new trucks were falling apart. Failure rates up to 40% were experienced. Since these heavy trucks were supposed to be durable for industrial jobs, the truck manufacturer's reputation was on the line. To complicate the situation, the first-tier supplier was secretly repairing adhesive bonds in the field without informing the warranty arm of the truck manufacturer. However, "things will out," and we found out. We calculated the actual loss to the truck manufacturer at $250,000 a year plus a large multiple for damage to reputation.

The most obvious solution, namely to change processes or to change suppliers, was complicated by contractual obligations and the time to renegotiate and plan, probably two years. The situation was so bleak that the truck company management had issued an edict (Manufacturing Feasibility Rejection) declaring the use of adhesively bonded SMC parts to be infeasible in manufactured products. The next step would have been an order to stop production, bringing heavy truck production to a screeching halt. The threat of this action was real and its implementation was rapidly approaching.

At that point in time, a nondestructive testing inspection method was recognized to be necessary. None was available. The truck company wanted to be able to inspect bonded truck bodies as they arrived at the assembly plant and to retrofit such inspection into the first tier supplier's plant. The truck manufacturing company wanted a field-portable method for obvious reasons.

The only test method available to the truck company at the time was a gross test for the absence of adhesive. A feeler gage shim was used as a probe between the two layers of SMC to detect whether adhesive was missing. This test proved ineffectual because many truck hoods were observed with the edges of adhesive joints "buttered over" with extra adhesive which prevented the entry of the shim. Sawing up these hoods revealed that the adhesive was missing from within the joints. Besides, the shim method did not address the question of weak bonds containing adhesive.

The plastics design group of the truck company assembled a task force and looked up as many NDT methods and instruments as they could find, but got no definitive answers off-the-shelf. They came to me as head of the NDT research, development, and applications group to evaluate these leads or invent a new method.

I put Gilbert B. Chapman, II, on the job and he singled out one suggested ultrasonic instrument as having some potential. This was the Sondicator Mk II manufactured at the time by Automation Industries and now redesigned by Zetek. The Sondicator used Lamb waves at approximately 25 kHz propagating between two closely-spaced probe tips. Actually, the wave motion involved both propagating waves and evanescent waves analogous to resonance near the tips. The received signal was compared in both amplitude and phase with the input signal by means of built-in circuitry, and poor bonds were signaled by a red light and an audible tone burst. The Sondicator required calibration against acceptable reference standards of adhesively bonded material.

The Sondicator was immediately found to be capable of detecting the difference between well-adhered adhesive in the lap joints and the lack of adhesive over moderate areas including "buttered-over" vacant regions. However, further work was required to detect the present but not-adhered adhesive and also adhesive with weak bond(s).

Chapman made a breakthrough on this question by making one important discovery, as follows. Namely, the Sondicator would reject almost all industrially made bonds if it was calibrated against perfectly made bonds in the laboratory. In reality, many of the industrially made bonds were strong enough to survive in the field. The test in this stage of development would have rejected all of production. Chapman's conclusion was that the "perfect" laboratory calibration standard was worthless. It followed that he had to create a calibration standard containing the requisite degree of imperfection to just barely accept the acceptable bonds and reject the bonds which were actually made but unacceptably weak.

Chapman solved the problem of the creation of sufficiently imperfect reference standards by applying statistics to a large family of bond samples made in the supplier's factory by hourly personnel under production conditions. These samples Chapman tested and rank-ordered with the Sondicator modified to give quantitative read-out, not just the red light and tone burst "no-go" alarm of its regular operation. Physical tensile pull-tests then determined the Sondicator level corresponding to the rejectable strength level. The reference standard was born as the type of sample just good enough to exceed the minimum specifications of the pull-test. With the reference standard, the "no-go" test could be used.

Chapman then taught the method at the plant where the trucks were assembled. The truck company also instructed the first-tier supplier on the use of the method and taught its own quality assurance surveillance agents to use the method so that high quality could be assured at the supplier and so that nonconforming product would not be shipped to the assembly plant.

The quality management office of the truck manufacturer accepted the method after Chapman wrote it up in the standard format. The method then served to define a specification for an adequate adhesive lap joint on a per-unit-length basis. No such specification had existed in the industry previously. The Chapman specification (Ford Motor Co., 1980) is now accepted as an exact parallel to the spot-weld specification for steel.

The edict declaring adhesively bonded SMC to be infeasible in a manufacturing context was rescinded just weeks before the order to stop truck production was to have been issued. One can imagine the magnitude of disruption which would have occurred if the company had been forced to revert to steel truck bodies. It would have impacted the plastics industry, the company's stamping plants, steel sheet orders, fuel economy, corrosion lifetimes of bodies, and all the future designs for a variety of SMC parts for further trucks and cars. As feasibility of adhesive bonding of SMC was reestablished, the use of SMC was extended to other parts and other car lines, thus improving corporate average fuel economy (CAFÉ) mileage and durability. The rescuing of SMC and the elimination of all the above problems is directly attributable to NDT applied with imagination and the requisite degree of smarts.

The cost of the NDT for keeping the SMC bonding process under surveillance for a year was about $25,000 including wages and the cost of the instrument. The first-tier SMC supplier reduced its failure rate from 40% to around 5% simply because it became cognizant that it could be monitored by the NDT "police function." Other parts went into production in later years because their bonding quality could be assured. NDT paid for itself many times over.

The method developed by Chapman is written up in his articles (Chapman 1982a, 1982b, 1983; Chapman et al., 1984.) The financial analysis is given in Papadakis (1985) and is used in one example in Chapter 9 of this book (see Section 9.2.2). A write-up of the scientific method as a nondestructive testing tool is given in Chapter 8 (Section 8.2.6).

Choosing bidders on price alone is bad, but doing so without methods to test their wares for latent defects is even worse.

Point 5, which follows, lies at the heart of Deming's manufacturing philosophy.

4.2.5 Point 5 Key Words: Improvements: Pinpointing

The decision made by executives in Point 1 is principally about improving quality after the idea of actually making the decision is absorbed. Deming's term for this, Continuous Improvement, has irreversibly entered the vocabulary of quality. However, the improvement must start with upper management because lack of quality entered the manufacturing system through management policies as shown in Chapter 2. Management must find more and more instances of the need for improvement over time as understanding improves, and must pinpoint the needed improvements.

This idea of Continuous Improvement is basic to the progress the Deming method expects to make through all the other points. Management created the problems under Taylor's tutelage and Ford's system; now management must solve the problems by using statistics to find their true nature and extent. *Special causes* of failures must be separated from *common causes.* Management should seek input from all levels of personnel including line, staff, labor, and consultants. All must participate in Continuous Improvement, according to the Deming plan. Labor may need to be empowered to participate in some solutions because the problems may have arisen through Taylor's elimination of the opportunity for labor to make a significant intellectual contribution. Note the earlier example of sparks in a welding machine in Chapter 2. Deming is trying to reverse the detrimental effects of having all knowledge and initiative kicked upstairs by Taylor and Ford in scientific management and mass production. Somehow the laborer must be enticed to become interested in quality once again after the loss of all his prerogatives.

In the bad old days, it was common for labor to chastise its own members for using their brains on production problems. I learned of one laborer who was making cams for cash registers around 1920. The laborers in this shop ground the curvature of the cam on a bench grinder one at a time by eyeballing it. The blanks had a square hole made previously by a punch press. This hole was intended to fit on a square shaft that connected the price key, by way of the cam, to the price sign to be pushed up into the window of the cash register. The laborer reporting his invention (Papadakis, 1975) told of putting ten blanks at a time on a piece of square rod stock and grinding all ten simultaneously. Needless to say, he outstripped the other laborers at piecework, and he came close to receiving a beating in the back alley. He not only earned more money but helped management. (Incidentally, he kept on using his intelligence and earned a Ph.D. in chemistry and became a professor emeritus in the end.)

The inspection technologist must question Continuous Improvement. This is not to say that there is any question about its long-term utility and, indeed, necessity. As it is usually explained, Continuous Improvement is carried out by calling the laborers together and holding a brainstorming session (Quality Circle) on the number of things that may have gone wrong. Sometimes a simple solution arises. Sometimes statistical work is instituted and results in the detection of special causes of problems. If the special cause needs a new, expensive piece of factory equipment for its elimination, then it may take two years to negotiate the purchase through the appropriations request process, budgeting, studies, bids, procurement, installation, and check-out. The possibility arises that the Continuous Improvement path as outlined may not be rapid enough to be classified as corrective action (see ISO-9000 in Chapter 5) to solve the problem. It may be predicted that inspection would be needed for a period of 1 year or 2 years while improvements are researched, developed, feasibility tested, and implemented. Inspection would have to pay for itself over that time period, assuming that feasibility of the improvement might be proved. Of course, it might not be proved feasible, so inspection might have to go on longer. This sort of contingency planning is not addressed by the Deming method.

Various other points have to do with human resources.

4.2.6 Point 6 Key Words: Training: Modern

While it might seem that the need for modern training would go without saying, one important aspect is stressed by the Deming method alone—training and empowerment of laborers to observe and fix problems. In particular, management needs to train the line operators to calculate and use statistics for control charts on the output of their machines. Then it must empower the operator to stop the production line if his machine goes out of control, and train the laborer to fix the problem if it is not too complicated; permit him, if necessary, to call an engineer or supervisor (as friend, not Tayloresque adversary) to fix complicated problems; and assure the laborers that supervision will commend them for improving quality, not condemn them for slowing production.

As a corollary, the trainers should be professionals in the field of training.

With training, the operators of the short-circuiting welding machine reported by Papadakis (2001) might have discovered the malfunction and precluded the need for two high-tech inspection engineers to do 2 weeks' work and then travel by company plane down to the factory in question. If a little knowledge is a dangerous thing, no knowledge is even more dangerous.

Lack of training can extend upward to engineers and designers, but on the other hand, their errors could have been caught on the factory floor by trained laborers. In this example (Papadakis, 2000a), trouble was detected in an automobile assembly plant when paint would not stick to car bodies. It was quickly discovered that a "PGM" tube had exploded in a hot primer bath, emitting silicone fluid (the PGM gel shock absorber in the tubular piston for the 5-mile-per-hour bumpers.) This PGM tube was examined and found to have an axial crack from the manufacturing of the tubing. During the manufacture of the PGM tubes from the raw material at a supplier, the original rod stock was turned down in diameter supposedly enough to get rid of all the manufacturing defects on its surface. The cylindrical hole down the centerline of this rod blank was pierced by forward extruding. Surface cracks could be exacerbated by the extrusion. As a final test, the PGM tubes were 100% inspected by an eddy current differential probe scanned over the entire surface by an automated machine. The author was called in by the automobile company as an expert in this NDT technology. The NDT system had been designed and installed as a turnkey operation by an NDT manufacturer believed to be reliable. The first thing the author observed was that the differential probe was not giving a failure signal on a test sample known to have a crack. The next observation was that the differential probe itself had a cylindrical shell and that this was mounted on the automated machinery inside a coaxial cylinder by means of a set screw. The set screw came loose, allowing the differential probe to rotate unhindered. Rotating the probe 90 degrees resulted in no signal because of the universal design of differential eddy current probes. Rotating was precisely what had happened. There was no flat for the set screw to seat against, defining the angle. There was no keyway to keep the angle constant.

The design engineer at the reputable NDT manufacturer should have been trained to put in a flat or a keyway. Even the ancient engineering symbol of the gear used as logo by Rotary International since its founding in 1905 has a keyway. These logo pictures are visible at restaurants and various other public places across the world for all to see. The NDT design engineer should never have turned out this eddy current probe holder design, and the applications engineer should never have installed it. The laborer should have been trained to note that lack of a signal from a "bad" part, inserted every morning as a check, was a sign of a malfunction. Some of Taylor's "kicking knowledge upstairs" even affects the best professionals. Forgetting lessons of the past is a dangerous proposition. The examples should have become second nature to the engineers to whom they were relevant. In a homely analogous example, widely distributed news reports from the tsunami of 2004 showed a case that was based on folklore, but important. Through oral history, people of one island remembered that circa 1900 a tsunami had come, first drawing the water level down in the bay before the onslaught of the incoming wave. From this oral history they knew that they should run for the hills if the bay went dry. In 2004 they saw the water recede and they ran. Only seven died instead of thousands.

In a less archaic vein, a law case concerning an airplane crash was settled by, among other things, proving that the designer of a new airplane knew or should have known of a certain safety feature built into a World War II airplane but left out of the design of the modern craft. Here is a description of one facet of the 1970 case from attorney Myron P. Papadakis, who at the time was assisting Houston attorney Wayne Fisher (M.P. Papadakis, 2005, personal communication).

From a system safety standpoint, the engineer is tasked to design his product with safety in mind. It is a well-quoted axiom that a system safety engineer designs out the hazards while the new widget is still in a paper and design prototype phase.

To help him in his judgments concerning the new widget he will utilize a 20–20 crystal ball, namely engineering experience, and tools of his discipline such as failure modes analysis, failure modes and effect analysis, fault tree analysis, and lessons learned.

It is far better to predict and eliminate hazard than to discover hazard as a result of an accident investigation. The experience in this case will demonstrate that fact.

Now fault tree as well as failure modes and effects studies are all, to an extent, based on supposition; lessons learned are as a result of understanding a historical failure or tragedy.

In the law, a manufacturer may be given latitude and some relief from extensive testing if the newly designed widget is substantially the same as an older one where testing was complete and safety seemed inherent. This precept is true for copycat drugs, for certification of aircraft and for many designs of most widgets.

The converse is the case when the widget is a departure from the state of the art (SOTA) or state of the industry (SOTI).

Now, as an example, if all we are going to do is switch an automobile from an aspirated engine to a fuel injected engine and by so doing achieve 10 extra horsepower, we may not have to test the entire vehicle again. Possibly only pollution emissions may need testing.

It is when you totally depart from the SOTI and attempt to introduce a new and radical design that you as a manufacturer have a duty of full testing and even unique testing. This new product requires stringent analysis and test.

Part of that duty to test includes researching the SOTA, which requires a look at Lessons Learned from previous but similar designs or applications.

Cessna, a manufacturer of General Aviation Aircraft, introduced a radical new aircraft in the mid 1960s. It was a twin engine, twin boom aircraft with high-mounted wings and retractable landing gear. Mounted facing forward was a centerline reciprocating engine.

Aft of the passenger compartment was a second, rearward-facing engine with a pusher propeller.

The wonderful simplicity of this aircraft as advertised by the manufacturer was the idea that if a general aviation pilot loses a wing-mounted engine on an ordinary twin-engined aircraft, the aircraft yaws terrifically at low takeoff speeds and a novice pilot would have his hands full.

Cessna advertised their plane with words similar to: The Cessna 337, Every man's P–38, Lose an engine, It is a piece of cake, with the center line mounting there is no yaw, so continue straight ahead like any single-engine airplane.

This seemed a good idea except that there were several incidents and accidents where the pilots had attempted takeoffs with failed rear engines. In the civilian design the engine instruments were not of optimum design or location and the pilot by design would not feel the loss of an engine with no yaw. Moreover, the location of the engine made it difficult to hear loss of power or see prop rotation stop.

In addition, some theorized that the rear engine housing design was such that engine failures due to air circulation and intake problems seemed greater in the rear than the front engine.

In our lawsuit we suggested that because of the poor instrument design and layout, and because of the inability of the pilot to see or feel the loss of a rear engine, he was unaware of his rear engine failure. We suggested that the airplane should be equipped with a rear-engine-out warning light. Our expert instrument designer's suggestion (an aviation psychologist from Wright Air Development Center, Dr. Walter Grether) was that the aircraft be equipped with a distinctive aural warning, a master red blinking caution light mounted in the straight-ahead cone of vision, and a red light within a feathering switch for the affected engine. Cessna maintained that this improvement was not needed.

I was on layover from flying an airline trip when I visited a bookstore in Ann Arbor, Michigan. It was there that I found a book with a picture of a Nazi fighter plane on the cover. It was a piston-powered Dornier 335 Pfeil (Anteater) aircraft. The amazing thing about this aircraft was the fact that it had one engine mounted in the nose and another pusher engine and propeller in the tail. As I picked the book up, I realized this was the only other centerline-mounted prop plane in existence. The United States shortly after the war had a half jet–half prop plane called the Ryan Fireball. This then was the genesis of the centerline thrust–low drag machine that Cessna was replicating. I paid for the book and took it back to the hotel.

To my amazement I read that a very early prototype of the Dornier 335 had crashed due to a test pilot's attempting a takeoff with a failed rear engine. It was a fatality. Nothing more was said about that pilot or that accident. I decided to find out what the state of the art was in 1942 and whether Cessna should have known.

I called the Smithsonian Air Museum and they said they indeed had the only Dornier 335 in existence, but that I better hurry because they were getting ready to ship it back to Dornier for a restoration and then it would reside in the Luftwaffe museum for ten years.

I called Adolph Galland—then president of the Luftwaffe Fighter Pilot's Association and the all-time world's leader fighter pilot ace. He placed me in contact with a former test pilot and I learned an amazing story about the aircraft. After the first fatal engine-out takeoff, the Nazis designed and subsequently installed an engine-out warning light called a Fuehrer Warning Lamp. It was installed in the cockpit for the pilot. Dornier in 1942 had learned the hard way what Cessna had not.

An interesting story—yes, but how did it tie into the manufacturer? As it turned out, after the war Cessna as part of the rebuilding process was to help Dornier re-enter the aviation marketplace. Cessna engineers were interfacing with Dornier people at their factories. I noted that the numbering system for the push-pull Cessnas seemed awfully coincidental. The Dornier number was 335 and Cessna chose the numbers 336 for their fixed gear push-pull aircraft and 337 for their retractable gear HUFF and PUFF. (The latter nomenclature developed as a slang name for the Cessna front-engine/rear-engine plane.) The numbers 336 and 337 were seemingly out of sequence for Cessna.

The case settled, and we suspect that a "Lesson" that should have been learned came back from a 1942 accident and reminded them to be ever vigilant in not forgetting "Lessons Learned."

(© 2005 Myron P. Papadakis. Unpublished. Used by permission.)

Modern training is certainly a necessity. Not only the training methods but also some of its subject matter must be modern. The subject of the training must be ancient as well as modern, reaching back to 1905 gears, 1930 statistics, and 1942 airplanes; and forward to ultrasonic, eddy current, x-ray, and nuclear probes. To keep up with modern methods, an in-house NDT engineering group is advisable for large companies.

4.2.7 Point 7 Key Words: Supervision: Modern

While this seems to be concerned with human relations and not technology, it is important for the technologist because it tries to unscramble the omelet Taylor made out of industrial labor and put Humpty Dumpty together again. Technology will work much better if labor is supervised correctly. The opposite of *modern* training is training that is domineering, adversarial, Theory X.

- The key is to be supportive, not adversarial. The author participated in one crucial case of this behavior. The human resources (HR) office accused one of my employees of malingering because he took every day of sick leave allotted to him every year. My boss leaned toward the HR position but gave me a chance to investigate. Upon questioning the employee, I discovered that he had a diabetic condition, which while under control, was serious enough to cause his doctor to classify his immune system as "brittle." The doctor had recommended that the man stay home and treat himself if he felt a cold coming on to prevent serious complications. He had been doing this, expending all his sick leave annually. I prevailed upon my employee to have his doctor prepare a letter for me spelling out in great detail the condition and the recommendations. When I presented the evidence to Human Resources, they backed down. The employee kept his job and kept performing well. Other personnel issues should also be treated equitably.
- Understand variability among people, day-to-day differences, morning person vs. night person, acrophobia (fear of heights), special problems. An example is divorce. Be extra understanding for a few months. One of my employees felt assured of my goodwill and actually asked me for patience and understanding for a while in just this circumstance. It is important for the supervisor to be patient, investigate the root causes as well as the symptoms of less-than-optimum performance, and find solutions that will help the employee perform well in the long run. As far as acrophobia is concerned, I could not walk along a catwalk with a low railing at the fourth floor level of a foundry. It was embarrassing, but I found another way to get from point A to point B.
- Investigate variable performance statistically and then worry only about the people who are out of control (i.e., who show outlying performance beyond three standard deviations). Seek to help, not to fire them. Determine what they may need, whether it be eyeglasses, machine repair, or whatever.
- From the point of view of statistics and averages, the following is the ultimate example: "Among all the Presidents, half are below average" (Deming, 1980).
- Make the following assumptions about people: If they are treated right, trained, and given a chance, they will put forth effort and do good work. This Deming positivism is the opposite of Taylor's negativism in assuming perpetual deliberate slowdowns.
- Treat people as if they are doing good work, and they will live up to the expectation so as not to ruin their reputations. (Example: One might consider giving ratings one step higher than deserved.)
- Give the supervised person as much knowledge and responsibility as possible (unlike Taylor). Certainly give him/her the responsibility

of running control charts for his machine and assuring the quality of its output.

- Enable the worker to have pride of workmanship. I repeat: even if he is just watching a machine, have him assure the output quality of the machine with a control chart so that he has ownership of the output. (Example: The senior engineer with the diabetes problem, mentioned above, was underutilized and undervalued. When I became supervisor of the group, I recognized this and gave him work at his level. He did well.)
- As foreman, supervisor, or manager, receive feedback from the worker and act upon it. Correct the indicated mistakes. Most mistakes are made by management, not labor, because management has sequestered all the thinking and planning.
- Commend, don't condemn, for all good intentions.

The modern methods of supervision along with some practical psychology are supposed to address the following:

4.2.8 Point 8 Key Words: Fear: Taboo

This is another human relations question that must be addressed by all management to undo Taylor's detrimental effects and Ford's stultifying system.

- According to Deming, fear on the part of workers is the greatest threat to good work.
- Fear is engendered by the Type X manager. Get rid of him or reprogram him. Teach him modern supervision techniques.
- Fear makes people unable to learn because they are afraid to look dumb by asking. They fear retribution, ultimately leading to termination. Jobs continue to be done wrong because the foreman does not know that the worker does not know how to do.
- Fear leads to defensive behavior and confrontations, which can lower productivity and quality.
- Fear is a source of fantasized future wrongs and a chip-on-the-shoulder attitude.
- Fear leads to "yes man" behavior.
- In a regime of fear, you have a "kill the messenger" approach, so there is no flow of information to correct errors.
- Managers who have all the answers engender fear because they are afraid to be contradicted by the truth from an underling. One manager I knew made it a practice to keep some negative evaluation for each of his people in his back pocket to be ready to use to take

the employee down a peg instead of building him up when the employee did something good. This manager always had a fabricated reason to explain why an employee did not deserve a higher rating or a compliment. This manager used fear. He once related how he told his son to imagine having a gun held to his head while he was studying for the SATs "to make him work hard."

- Downsizing is the newest killer of productivity because it assures a continuous atmosphere of fear where good work is no longer rewarded. (This phenomenon came after Deming, so he did not address it.)

4.2.9 Point 9 Key Words: Teams, Not Barriers

Taylor set up barriers with his organization charts and job descriptions. All forward motion of a plan had to go "over the wall" to the next department. Barriers are visualized as walls in this construct. It would be better to organize interdisciplinary teams to do concurrent engineering rather than to have individual specialties going over the wall, over and over. This effort helps technologists, including NDT experts, impact the operation of the company without delays, red tape, and turf wars.

- The tendency of each area "Before Deming," was to engage in "empire building" without concern for the entire company. Without coordination, each area suboptimizes itself with gross added costs to the company.
- Over the wall mentality leads to major rework cost. Over the wall means that each area finishes its work and then tosses the result with plans and specifications "over the wall" to the next area to use or implement. Modern parlance also talks of each operational area being in a "chimney" where there is no contact between one group and another except when finished designs are passed forward. Interestingly, this compartmentalization is spoken of as a virtue in the Taylor method. Parkhurst praises the over-the-wall practice (not in so many words, but the image is exact) in describing the reorganization of the medium-sized manufacturing company for which he consulted (Parkhurst, 1917, 8–9 and Figure 1.)
- In a company ruled by the over-the-wall mentality, the first area has no idea what the next area (its customer) needs.
- The following things are done in sequence in a manufacturing firm dominated by over-the-wall mentality:
  - Marketing perceives a customer desire. They call in the research department.
  - Research gets a novel idea on how to make the item that marketing has suggested. After some investigation, research tosses the idea to design engineering.

  - Design develops a design and tosses it to product engineering.
  - Product engineering makes plans for a realizable gizmo. It then tosses the plans to manufacturing and engineering.
  - Manufacturing makes plans for the processes needed to build these things. The plans are then tossed over the wall to industrial engineering.
  - Industrial plans a factory and turns responsibility over to plant engineering.
  - Plant builds or modifies a factory and turns it over to production engineering.
  - Production is faced with the day-to-day task of building the item.
  - Quality control (QC) and NDT are called in as needed with no preparation.
  - Finally the user (ordinarily called the *customer*) receives it.
  - At some point, maintenance is needed.
- Without input from the next stage (the immediate customer), there is tremendous waste due to one stage finding it impossible to implement the ideas of the previous stage. One engineer years ago phrased it this way: "Architects draw things that can't be built" (Eastman, 1947, private conversation).
- Change orders, deviations, rework, redesign, and so forth ensue.
- The antidotes for this situation as suggested by W. E. Deming are:
  - Product line teams should be instituted instead of professional areas. (Examples of product line teams are the modern Chrysler and Ford organizations for car platforms, and the Lockheed Skunk Works for spy planes.)
  - Concurrent engineering (simultaneous engineering) should be used throughout. Get together a team from all areas (including marketing) starting on the day marketing suggests a new product. Work on all aspects from the beginning. Ensure cooperation and no surprises. Be prepared by inventing inspection methods for new materials and structures.

One classic example of the need for 100% inspection to fix an over-the-wall problem arose in the automobile industry. The need was recognized after the supplier's protracted efforts at Continuous Improvement. Indeed, the supplier organization averred that it was producing 100% conforming materials. As a backup, the automobile company had involved NDT in its concurrent engineering of the part in question. NDT saved the entire product line, which was to be a new compact car desperately needed during an oil embargo. The short report on the NDT as part of the concurrent engineering process is reproduced here (Papadakis, 2002, 1292–1293).

There is nothing more basic in NDT than having a test ready when it is needed. This Back to Basics article is a case history of preparing a test and finally getting it implemented. "Finally" is the right word, because the test was rejected by upper management until the night before Job 1.

"Job 1" is automotive jargon for producing the first item in the factory where the items will be produced on the day production is begun on a new item. All the equipment is in place, all the hourly workers are at their stations, all the raw materials are on hand, and the pistol shot is fired to start the race, figuratively.

The new part in question was a powder metal connecting rod for a new I4 engine. That is a four-cylinder in-line gasoline automobile engine. The new engine was to power a million new-model compact cars in the following 12 months. The profit on those vehicles hinged upon the success of the powder metal connecting rods.

Connecting rods connect the pistons to the crankshafts. The rods take all the stress of the fast-burning gas mixture on the power stroke and of the gas/air being compressed on the up stroke. "Throwing a rod" can destroy an engine.

Originally, connecting rods were made of steel forged at red heat. Some rods were later made of nodular cast iron. Powder metal was envisioned as a strong and economical substitute for both.

Powder metal parts start literally as powdered metal, which is compressed into a mold to form a "pre-form," which is then sintered to become a solid metal. For adequate strength, the piece must be "coined," which means compressed further at high temperature in a tool and die set to give the final shape of the part. Only a minimal amount of machining is done after the coining of the near-net-shape part.

Research and development was begun more than two years before Job 1 in the Manufacturing Processes Laboratory at the automobile firm. At two years before Job 1, my NDT group was called in to join the concurrent engineering team working on the powder metal connecting rods. The chief metallurgist told us that several potential failure modes of the powder and the process had been discovered, and that the engineers needed NDT methods to detect these failures, should they occur in production. The failure modes included oxidation of the powder, wrong composition of the powder, inadequate filling of the pre-form mold, cracks in the fragile preforms before sintering, and improper temperatures. The chief metallurgist told us further that the failure of one rod in 10,000 could bankrupt the company. (This was in the hard times after the second oil embargo in 1983.) NDT was under the gun.

A scientist and an engineer in my group went to work on the problem using specimens deliberately made to exhibit these defects by the chief metallurgist's staff. A low-frequency continuous wave eddy current method was developed which was capable of sorting each type of defective specimen from the acceptable specimens. This method was written up and turned over to engine division for implementation at an appropriate location before the first machining step. The technology transfer occurred more than a year before Job 1. We were prepared. My NDT group went on to other projects.

A few days before Job 1, the coined parts began arriving from the powder metal processing specialty supplier. The chief metallurgist made a quick run down to the engine plant, picked out a few coined parts at random, and ran metallographic tests on them to satisfy himself of the quality. By this time he was officially out of the loop, but he wanted independent confirmation that "his baby" was going to be born okay. And what did he find? Precisely the metallurgical problems he had predicted in the failure mode analysis! He blew the whistle and got the attention of executives up to the vice-presidential level. My NDT group was called in because we had the solution. But why had it not been used?

This emergency was the first time we had heard of the actual production scenario that had been decided upon by engine division. They had decided to outsource the powder metal parts to a specialty house which would take care of everything between the design which the auto company supplied and the delivery of the coined part. They had claimed that they could produce everything perfectly. They averred that NDT would be unnecessary. Engine division bought off on this assertion and did not call out the implementation of NDT. Their error was discovered by the diligent chief metallurgist just hours before production of garbage was to commence. The error could have led to hundreds of brand new cars throwing connecting rods on interstates.

A series of high-level meetings was held. I had the opportunity to explain our NDT method made available by concurrent engineering a year ahead of time. I enjoyed watching the auto executives force the powder metal specialty house to back down, swallow their words, and install my NDT. To bring about the implementation, I had to lend engine division two eddy current instruments with coils, my group's whole complement of eddy current gear. One was used in the engine plant to sort the 60,000 parts already delivered. The other went directly to the powder metal specialty house, and the one at the engine plant ended up there, too, after the initial sorting. They were forced to buy their own as soon as delivery could be arranged.

Job 1 on the connecting rods, the new engine, and the advanced car were all saved by concurrent engineering including NDT. If concurrent engineering had omitted NDT, then Job 1 would have been delayed a few weeks until an NDT test could have been developed on an *ad hoc* basis. Imagine, if you will, the loss from shutting down an engine production line and a car production line, each scheduled to run 60 units per hour for two 10-hour shifts, for three weeks. If the planned profit were $5000 per car, then the loss would be 108 million dollars. That is penny-wise and pound-foolish if you consider the cost of two hourly workers and two ECT instruments at $4000 each.

So what is basic in this lesson? First, you need the scientist and the engineer to invent the test that **will become basic** a few days or weeks (or even years) in the future. Second, you need to **involve NDT up front** and not call upon it as a last-ditch effort. Drain the swamp. Preempt the alligators. Third, **do the failure modes and effects analyses** to find out what tests you may need to generate with your concurrent engineering. Finally, don't let any smooth-talking snake oil salesmen tell you that NDT is not needed.

4.2.10 Point 10 Key Words: Slogans: Counterproductive

While this issue deals primarily with interpersonal relations, there is a big dose of keeping processes in the factory under control here. The term *slogan* is a catchall for harangues, irrational targets, browbeating with cute pictures, and so on. Every worker has his own pet peeve.

- Targets and slogans are counterproductive. Management should eliminate targets, slogans, pictures, posters, and so forth from the workplace, thus urging the workforce to be more productive.
- Do not force workers to sign their work. "Box made by Jack's team" and "Inspected by No. 17" are examples to be discouraged. The work is done exactly as required by the management by means of machines, and forcing the laborer to sign off is insulting to him or her.
- It is management's job to ensure that all conditions are under control and the best available so that the worker can do a good job all the time.
- Exhortations are not needed; instead, management needs a plan for improvement. Providing and implementing a plan is a requirement of management.
- Examples:
  - "Zero Defects" is *prima facie* impossible and is an ad hominem insult to workers.
  - "Do It Right the First Time" is another backhanded slap at workers who would do this naturally if management would give them the right conditions, tools, and respect.

- Charts on the wall showing that workers have not yet met the artificially high goals set by management are counterproductive. Again, earnest workmen are insulted and look at this as Taylor's "hustling."

4.2.11 Point 11 Key Words: Quotas: Taboo

While this may look the same as the previous point, it is quite different. It involves keeping things statistically under control for the workers. First we need some definitions. As one will realize, quotas are the same as work standards when this means number of parts per hour and so forth. The laborer should be paid for the hours he puts in, and the work output should be arranged by the management through the adjustment of machines to let the worker do good work at a reasonable rate with the process under control. While the process is under control, the production of nonconforming material is statistical and is not the laborer's fault. Management by objectives (MBO) is taboo also, as one can game the system.

- Using *work standards* means that
  - You (the laborer) have to produce a quota of parts in a day.
  - You may not produce more than a certain number of defective parts per day.
  - What are the consequences?
- This measure leads to
  - Despair among honest workers when the conditions, materials, machines, and methods are not adequate (management is at fault).
  - Shipping only the quota even if more could be made. This inhibits progress. (Example: laborer who made brace of ten cams at a time, above.)
  - Shipping bad parts (failed inspection) after faking the QC records to fulfill the production quota. (See the example involving auto crankshafts, Kovacs, 1980, personal communication.)
  - Shipping bad parts so they will not be charged against your number of defectives.
  - All these are bad business and bad motivation.
- The true remedy is as follows:
  - Management should set up a production system with a known process capability (including the Four Ms: men, materials, methods, machines, and environment. See Figure 3.1). This means that the output per day and the defectives per day will be known statistically, *a priori*.
  - Start it out under control (statistically).

  - Train the worker.
  - Empower the worker to keep the process under control with control charts (i.e., to detect the time when the process goes out of control by using control charts as the means of detection).
  - Simply accept the statistical fluctuation of output and defectives. (They will be only outliers while under control.)
  - Invent and install 100% electronic inspection if the process capability cannot achieve few enough nonconforming parts.
  - Accept what the workers can do if it varies from worker to worker. (Example: Suppose you have 20 workers; is the performance of any beyond 3 standard deviations? With good modern supervision, find out the reason why.)
- Management by objectives is bad, like quotas.
  - The objectives may be understated. For example, a worker may keep accomplishments in his hip pocket like a just-in-case inventory of accomplishments to quote later as needed at a time of poor performance.
  - The objectives may be overstated. For example, they may be imposed by the manager, and be unrealizable.
  - MBO creates fear of underachieving. Fear is counterproductive, as in Point 8.
- A point to remember: Everyone will do his best if treated right. Labor needs loyalty down as much as management needs loyalty up. Note that this is a Deming belief contrary to the Taylor belief in lazy men and *soldiering*, which meant never volunteer, never do more than the minimum, never go over the top unless ordered.

4.2.12 Point 12 Key Words: Workmanship: Pride (Remove Barriers That Hinder the Hourly Worker)

The items in Points 10 and 11 on slogans and quotas, as well as Point 8 about fear, all contribute to Point 12. Many of the points stated positively previously, are reiterated in the negative here for emphasis. Genuine pride of workmanship should be enhanced. In the present milieu where the worker is no longer the master with self-determination, this is hard for management to accomplish. Management must attempt to eliminate all barriers to the pride of workmanship.

All barriers do one basic thing—take away the pride of workmanship. How can a worker have pride of workmanship if

- Management supplies him with junk input so his efforts do not produce good output?
- Management does not keep machinery in good repair so even good effort has bad results?

- Management does not keep gages in repair to tell a worker whether his output is or is not any good? (Note the entire section on calibration of gages in the ISO-9000 standard.)
- Management does not provide training so that the worker can even know what to do?
- Foremen just want production, and quality takes a back seat?
- Management does not listen to the worker's suggestions even though he is saying something as fundamental as "Hey, guy, this machine is crapping out" (Taylor theory says management should not listen)?
- Management retains fear as the principle of management?
- Management views labor as a commodity in an economic equation where the solution is move to mainland China and fire all American workers?

Contrary to this, Deming believed "the performance of management... is measured by the aim to stay in business, to protect investment, to earn dividends, and to ensure jobs and more jobs.... It is no longer socially acceptable performance to lose market and to dump hourly workers on the heap of unemployed," (Deming, 1982, i).

TQM seems to have a fundamental contradiction: If you cannot stay in business unless you fire all your American workers, then you cannot treat your workers decently as TQM requires.

4.2.13 Point 13 Key Words: Education and Training

Assuming that a company has decided to make the commitment to stay in business with American workers (as Deming was speaking of American companies and American workers), then a great deal of training will be needed. The Taylor emphasis of having employees with no knowledge walking the floors of the plant will have to be reversed.

- The company must teach statistics so everyone can learn how to do the following:
  - Manage (given the variations in people and performance)
  - Design (using factorial experiments, statistical dimensioning, etc.)
  - Produce (using control charts to indicate when processes go out of control)
  - Choose or reject 100% inspection
- There will, of necessity, be changes of field for many workers and technologists.
  - Fewer quality control functionaries, more statisticians
  - Fewer routine tests, more hi-tech monitoring and auditing (see ISO-9000 in Chapter 5)

- It is necessary to introduce a new emphasis and reeducation for management.
  - Theory X managers must be reprogrammed.
  - Everyone (management and specialists) must take four days of Deming lectures (this chapter's TQM, expanded) and five days of a specialty. The trainers must have modern training (Point 6).
  - Renounce belief in or adherence to the doctrines and teachings of other quality gurus. Adhere to Deming alone as taught in Points 1 and 2.

4.2.14 Point 14 Key Words: Implementation: Staffing

All of the above must be implemented with executives and management giving 110%, as they say. To do this, executives should create a structure in top management that will work daily to accomplish the first 13 points. The structure must be staffed with the proper experts, who must be given authority to insist upon Deming-consistent performance from all other staff and line areas of the company. The structure must reflect all of the following:

- Responsibility
- Authority
- Power
- Budget
- Personnel
- Expertise, including SPC at all levels
- Belief in TQM as emanating from the CEO, president, and stockholders
- CEO must force actions by vice presidents (VPs) in conformity with this new staff structure

Examples of errors:

- One large multinational corporation with many executive VPs, VPs, executive directors, directors, and lesser personnel made the head of the Deming change brigade only a director.
- One director heading up an independent unit in an organization, professing to implement TQM, hired an organizational development manager and assigned him responsibility but delegated no authority, gave him no power, provided no budget, and on top of that, practiced nanomanagement (worse than micromanagement) by being a know-it-all, not letting the manager be a self-starter with ideas, and even censoring outgoing mail.

- An executive director in a multinational company, convinced that he could achieve total manufacturing perfection (zero defects) without statistics or feedback, decided to rely on deterministic technology and forced his manufacturing committee to write a white paper advocating this.

4.3 Summary

In this chapter the way SPC and statistical thinking, to use Deming's terminology, are integrated into TQM has been outlined. It has been pointed out that 100% inspection plays a role in certain processes when SPC is in place. The proof of the financial utility of inspection will come in Chapters 7 and 9. Deming-approved personnel have downplayed the ideas of inspection of 100% of production of any part on the basis of adherence to Points 3 and 5—the ideas that inspection is taboo and that improvements pursued rigorously over time will always be adequate to preclude the need for inspection.

The idea of proving mathematically, as I shall in this book, that inspection can improve profits is basically anathema to the statisticians among quality professionals. In the later chapters it will be proved that inspection can make a profit. Inspection is the proper course of action in several classes of manufacturing problems.

5

ISO-9000 with Statistics and Inspection

5.1 Background

The International Standardization Organization (ISO)-9000 quality management standard is the document around which modern quality is managed (ISO 1994a, b; ISO 2000). It is an evolving document first issued in 1990, with a version 2000 now available. ISO-9000 contains the ruling system for the management of quality in all countries that need to trade with the European Economic Community. ISO-9000 provides the set of priorities by which quality is managed, and is arranged by topics and subtopics. In this text the 1990 version will be studied first, then its evolution will be traced. The inspection technologist (nondestructive testing [NDT] expert) should become familiar with the quality standard. The first few topics are the most important, but the remainder should also be studied. The 1990 document comes in three versions, 9001, 9002, and 9003, for companies with progressively less involvement in the high-tech aspects of a business. ISO-9001 covers design and development functions as well as production and all the rest, while 9002 does not include design and development. Many companies produce a product from the designs of others, so a lower tier of documentation is useful to them.

The standard makes these assumptions:

- Quality must be managed; it does not just happen.
- There are various options for managing quality.
- It is necessary to standardize using one method.

The origin of the standard is the ISO, which wrote it. The United States had some input through the National Institutes of Standards and Technology, but the document is basically a European Economic Community (EEC) paper. The standard was advocated first by the EEC as a way to bar from trade any goods not up to their standards. They had set 1993 as a cutoff date to blackball goods if companies did not achieve ISO certification. Early on, some viewed this as an attempt at creating a European cartel, but through international registrars (firms authorized to grant certification under exacting conditions) non-Europeans have been admitted. Companies in countries in the Western Hemisphere, Asia, and elsewhere have achieved and maintained certification.

The table of contents of Part 4 of the ISO-9000 quality management standard (1990), the substantive operational part, was shown in Table 1.2. This is simply the list of subjects covered that are relevant to quality. The reader is requested to refer back to that table as needed. As can be seen, the document is a general system for quality management. It gives generalized rules addressing all activities of an organization that may have an impact on quality. Specifically, it does not single out any quality department or quality control manager or vice president for quality. However, it mandates many requirements pertaining to the rules it does set down. Registrars periodically audit based on these rules to permit the organization to maintain its certification. The verbiage is of a general nature allowing application of the requirements to all organizations. Specifically, the standard does not cover any aspect of operating a specific business or a class of businesses. For example, it does not cover

- What process to use
- What materials to use
- What accuracy to require
- What measurements to make
- What instruments to use
- What data to take
- What form to record it in (e.g., electronic or hard copy)
- How long to retain the data

However, the standard has the following requirements on documents and their utilization. Please refer back to Table 1.2 for the sections of the standard. Also note the Five Tiers of Quality Management listed in Section 1.3 of Chapter 1. We will begin with the 1990 version.

The standard requires that a quality manual (a document) be written to address each of the 20 sections of the standard. It requires further that the quality manual refer to written procedures for each activity carried on in the organization and to the recording of the results for permanent quality records. It also requires that each written procedure refer to written work instructions for the floor personnel to follow to do the job and record the results. These record forms are documents, also. The work instructions arise directly from Taylor's scientific management.

Another requirement, seemingly obvious, is that all documents for current use be up-to-date versions. Anything seemingly obvious must be written down. Beyond this, there is a requirement that management make sure that all of this happens.

So, what do the originators of ISO-9000 believe will happen if the ISO-9000 quality management standard is implemented?

5.2 ISO-9000: Keeping a Company under Control

The basic philosophy behind the ISO-9000 (1990 version) is that if you have produced good quality in the past, and if you know what you did, and if you continue to do it, then you will continue to produce good quality in the future.

Conformance to ISO-9000–1990 ensures this continuity through *documents* establishing what you need to do, through *records* to show that you did it, and through *audits* to assure your customers that you actually conformed.

ISO-9000–1990 also provides methods for fixing things if they go wrong by preventative and corrective action, management reviews, and instructions on data gathering. ISO-9000–1990 also gives you methods and reminders on how to approach novel occurrences such as design control and concurrent engineering for process control of new items (in Level 9001).

Beyond this, you should get your organization under control by writing down all aspects of your methods of doing business, by making additions to this formulation to cover the 20 sections of the standard, and by achieving ISO-9000 certification to (a) inform your customers of how good you are, and (b) discipline your organization. You should keep your organization under control by performing according to the documents, by recording this performance, and by reviewing and auditing the documents and the performance by means of (a) management reviews, (b) internal audits, and (c) external audits to maintain certification. The result, according to ISO ideas as of 1990, will be high quality.

Certain industries have decided to make formal additions to the ISO-9000 standard to make it industry-specific. It is not the purpose of this book to go into all of these permutations of the basic standard. The practicing engineer will become familiar with his or her own industry-specific standards.

With all the emphasis on statistics in Chapters 3 and 4, what does ISO-9000 say about statistics?

5.3 Statistical Process Control and Statistics within ISO Philosophy in the 1990 Version

Comparing the way statistical process control (SPC) and statistics in general permeate total quality management (TQM) and the optional way they are treated in the 1990 version of ISO-9000, it is probable that the TQM advocate would suggest that without statistics, the company planning to adopt the ISO quality management standard could only have produced its high quality to date by accident, and that it would be simply accidental if the ISO system seemed to maintain high-quality production in the future.

While statistics is mentioned in the ISO-9000–1990 standards, the language is not imperative in the sense that the verb phrase *shall use statistics* is not employed. Rather, the standard says that the supplier must look into the possible use of statistics and then must document the use of any statistics he decides upon. This means that the supplier (your company) can come to the conclusion that no statistical procedures are needed, and declare such a verdict. The assertion cannot be questioned if a corporate officer of high enough rank will sign off on it so that it is entered into the quality manual. This outlook is directly contrary to the TQM idea of statistics. (Note below that version 2000 of the quality management standard is stricter on statistics.)

If statistics was downplayed in the 1990 version of the quality management standard, how was inspection treated? Is NDT in or out?

5.4 Inspection in ISO-9000–1990

One might ask where NDT fits into the control of quality. The answer is—everywhere. Symbolically, the possible interjection of NDT and inspection in general into quality systems is shown in Figure 5.1. Inspection, particularly NDT, within ISO-9000 is interpreted in this section.

ISO-9000–1990 actually calls out inspection in several sections. The operative phraseology is, "Inspection shall be carried out…". This specificity is in direct contradiction to TQM, which eschews inspection except as an exception limited in application and duration. ISO-9000–1990 specifies inspection forever in several situations of general scope. These uses of

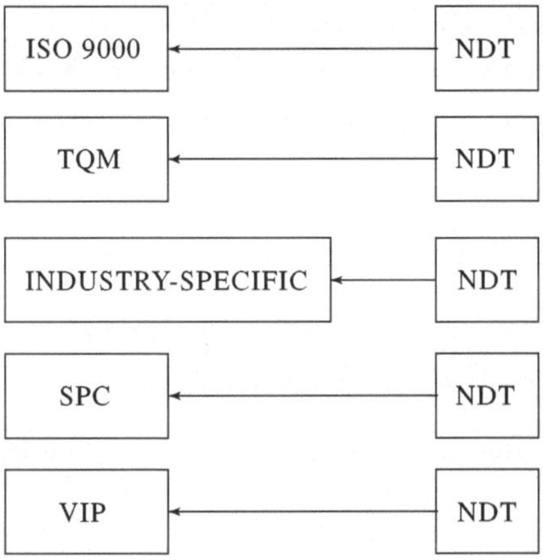

FIGURE 5.1
Symbolic diagram of the fit of NDT and inspection in general into the kinds of quality systems in existence. ISO-9000 and TQM are explained at length in this book. Industry-specific systems are add-ons to ISO-9000. SPC has been explained to the degree necessary. VIP is verification-in-process, a procedure for putting inspection into the production line at critical locations.

inspection are pointed out and studied here. In the sections specifying inspection, only the relevant subsections are noted. Subsections not mentioning inspection are not referenced for brevity. Reference is made to Table 1.2 in this book.

The first section, management responsibility, is general and does not say anything specific about inspection. The second section, quality system (quality manual), makes several specific references to inspection and its other manifestations, such as testing. The relevant passages mean the following:

Section 4.2 on the quality system says that the organization must carry out quality planning in several areas. One important area is to keep abreast of the state of the art in testing techniques for quality control and inspection. Equipment is to be updated as needed. The organization should even plan to develop new inspection instruments if it identifies a need. A second area of planning is to pinpoint needs relevant to future products. If inspection instruments are identified as unavailable, the organization should undertake research and development in a timely fashion, possibly years ahead of time. Third, the planning effort should extend to instruments needed for verification-in-process to ensure good output.

Thus, thinking about the need for inspection is embedded in the quality manual at the heart of the 1990 version of the 9001 full text of ISO-9000. One is supposed to think about acquiring state-of-the-industry equipment, of course, but also one is supposed to plan to develop new state-of-the-art equipment in time for use when new processes come online. One's staff is supposed to think ahead, to define the need for new equipment, and to develop it. Use of concurrent engineering involving NDT engineers and other electronics experts, along with the process and product development experts, is implicit in this directive. The example of powder metal connecting rods given earlier in Point 9 of Chapter 4, Section 4.2.9 epitomizes the efficacy of the directive in this section of ISO-9001. In that example, an NDT inspection method was developed 2 years in advance of the time it was needed. Concurrent engineering ensured this good result.

After this, the useful text jumps down to Section 4.9 of ISO-9001, which is entitled process control. The organization is supposed to carry out all its processes under controlled conditions. Among the process control regimens specified are monitoring of process variables and product characteristics. Both types of monitoring require instruments that should have been investigated by the planning function above. In particular, product characteristics that depend on intrinsic physical variables or latent defects must be monitored by NDT, although NDT is not mentioned specifically in the standard.

Processes must be monitored for their own parameters such as time, temperature, humidity, pressure (force), voltage, amperage, and so on. Processes turn out products as shown schematically in Figure 3.1, so the products must be monitored to see that the processes actually had their desired effects. Because the desired effects may be characterized by either extrinsic variables or intrinsic variables or both, appropriate methods to monitor such variables must be used.

It is particularly important to monitor intrinsic variables because improper values of intrinsic variables are often causes of latent defects. Frequently it is possible to use NDT to detect improper values of intrinsic variables by correlations. Research is needed to establish the correlation between the value of an NDT parameter and value of the intrinsic variable that one wishes to measure. See Chapter 8, Section 8.3, of this book. After the research has established a curve with error bands, the NDT parameter may be measured on production parts, and the value of the intrinsic variable can be predicted within the empirical errors. The electronic NDT methods are rapid and cheap, neither interfering with production nor increasing its cost (except incrementally) even if 100% inspection is needed. Several such electronic NDT methods will be explained in Chapter 8, Section 8.2.

It is important that the monitoring of product characteristics is addressed in ISO-9001. The International Standardization Organization recognized the advantage of doing product monitoring.

The next section, 4.10, is explicitly about inspection and testing. Incoming, in-process, and outgoing inspection are specified. They must be documented in the quality plan and procedures, and records of their performance must be kept. Receiving inspection is supposed to be performed before the raw material enters into a process. Before product is released from one workstation to another, the in-process inspection must be performed and documented. Some deviations within the factory are permitted. No product is to leave the factory until all the testing procedures are performed and recorded. This includes all final inspection and testing as well as all the rectification of incoming inspection deviations and in-process deviations.

It is important to note that the 1990 version of ISO-9001 insisted upon incoming, in-process, and outgoing inspection of raw materials and product. The absolute prohibition on the shipment of material before it had passed all its stepwise tests is a firm acknowledgment of the need to inspect a product. The use of NDT to inspect for latent defects arising from intrinsic variables is something to be considered within this context.

The next section, which the reader may interpret as suggesting that inspection might be appropriate, is Section 4.14 on corrective and preventative action. While the word *inspection* is not used, one can think of scenarios in which inspection of 100% of production might be decided upon to eliminate the causes of actual or potential nonconformities in shipped product. Doing inspection commensurate with the risks might be a rational procedure.

Using 100% inspection by NDT or other scientific techniques to eliminate as many nonconforming parts as possible, consistent with the probability of detection of the method, might be commensurate with the risks encountered. SPC might not eliminate as many. Certainly inspection would be at least a line of last resort while other approaches were being investigated. If the causes of the nonconformities could be addressed by 100% incoming inspection or by 100% inspection for verification-in-process, then its use would be eminently reasonable.

The organization's response to these sections of the standard can be analyzed further as follows.

- A possible response of a firm to many of these clauses could be to write down that no action is to be taken. This response must be written down, justified, and signed off on, because an auditor can ask the question, "What do you plan to do if…?" to satisfy the standard.
- No specific branch of math, science, psychology, or engineering is referenced anywhere in the standard, which is a generalized quality system standard.
- On the other hand, the development of new inspection instruments where needed to meet current and future production needs is mandated. The onus is on the organization to prove or at least assert at a high level of responsibility that such will not be needed.
- However, it can be argued in some cases that NDT is the most appropriate approach. NDT could fulfill the requirement in 4.14.1, for instance.

This book will give financial calculations showing that NDT is appropriate to counter the risks in many cases.

As we have said, ISO-9000 is a living document. How has it matured and changed going into the year 2000 version?

5.5 Changes in Emphasis in the ISO-9000–2000 Version

The changes can be stated up front as follows: *continuous improvement* has been added. Other changes include the requirement for the use of statistics, the elimination of requirements for the use of inspection, and the addition of requirements concerning recognition of customers' opinions. The order of presentation has been changed so that the 20 topics in Table 1.2 are now grouped under five categories. One draws the logical inference that quality professionals versed in TQM have influenced the committees writing the ISO quality standard. Time will have to judge the efficacy of the new standard. In detail, the changes begin with continuous improvement. We look first at the philosophy of change.

5.5.1 Philosophy

The ISO-9000–2000 standard is based on the idea that continual improvement is necessary in addition to continuing to do good work and documenting it. If every organization is continuously improving and yours is not, then you will fall behind in quality. Continual improvement is playing catch-up in advance. The ISO-9000–2000 standard also assumes that the customer's expectations for quality are continually rising and should be incorporated into an organization's forward planning.

The reorganization by categories is outlined in the following section.

5.5.2 Reorganization

- The ISO-9000–1990 standard gave general categories of activities related to quality but in a less logical order than the 2000 version.
- The 1990 version of the standard (Table 1.2 in Chapter 1) contains 20 categories (sections) in its Part 4, whereas the 2000 version is organized under five categories (sections).
- The activities from 1990 are all included in the 2000 standard by reorganizing them under the five new categories (sections).
- Nomenclature: *Supplier* is now someone your organization buys from; you are the *organization* and your organization has *customers.* This nomenclature is more nearly consistent with the vocabulary used in most industries.

The additions to the quality management standard for 2000 are specified in the next section.

5.5.3 Additions

- The ISO-9000–2000 standard has introduced two new ideas as requirements:
  - Customer orientation
  - Continual improvement
- These ideas are interspersed within the five sections, as will be shown shortly.
- Customer orientation appears at both ends of the design-to-sales sequence, namely
  - As customer input to determine the characteristics of a high-quality object, analogous to Table 1.1 of this book
  - As feedback to determine whether the customers think the produced object meets their quality expectations
- Continual improvement appears throughout. Everything is to be improved including the organization's quality management system (within the context of ISO-9000–2000). Improving the quality management system requires proactive management.

The next section discusses how the standard treats the three levels of business as applied to organizations.

5.5.4 Applied to Organizations

The ISO-9000–1990 standard had three operational levels, namely 9001, 9002, and 9003 as listed previously.

The ISO-9000–2000 standard has only one level, 9001, which can be used with specified deletions for the organizations with less extensive scope. In other words, it is used by exception or deviation for simpler organizations.

Again, as with ISO-9000–1990, the first three sections in ISO-9000–2000 are introductory material. The substantive standard is in the five sections numbered 4 through 8. Their table of contents is given in Table 5.1.

Essentially all of the material in the 20 sections of Unit 4 of the 1990 standard is rearranged into Sections 4 through 8 of the 2000 standard shown in the table. The new material on customer orientation and continual improvement is sandwiched into these five operating sections. These and other changes will be explained.

TABLE 5.1

Table of Contents of ISO-9000–2000

1, 2, and 3: Introductory Material
4. Quality Management System
4.1 General Requirements
4.2 Documentation Requirements
5. Management Responsibility
5.1 Management Commitment
5.2 Customer Focus
5.3 Quality Policy
5.4 Planning
5.5 Responsibility, Authority, and Communications
5.6 Management Review
6. Resource Management
6.1 Provision of Resources
6.2 Human Resources
6.3 Infrastructure
6.4 Work Environment
7. Product Realization
7.1 Planning of Product Realization
7.2 Customer-Related Processes
7.3 Design and Development
7.4 Purchasing
7.5 Production and Service Provision
7.6 Preservation of Product
8. Measurement, Analysis, and Improvement
8.1 General
8.2 Monitoring and Measurement
8.3 Control of Nonconforming Product
8.4 Analysis of Data
8.5 Improvement

5.6 Overview of Sections 4 through 8

5.6.1 Section 4: Quality Management System

The first paragraph in Section 4 provides general requirements and states that the organization must proactively analyze itself, draw up a quality management system according to the ISO-9001–2000 standard to fit its needs, and implement the system. Parts of Sections 4.1 and 4.2 of the 1990 version are analogous. The second paragraph in Section 4 addresses the whole area of documents and covers the remainder of Section 4.2 plus Sections 4.5 and 4.16 of the 1990 version.

5.6.2 Section 5: Management Responsibility

The first paragraph in Section 5 covers part of 4.1 on management responsibility in the 1990 version. The second paragraph in Section 5 is on customer focus and is new. The third paragraph in Section 5, called quality policy, covers portions of 4.1 and 4.2 in the 1990 version. The fourth paragraph in Section 5, called planning, also contains portions of 4.1 and 4.2 in the 1990 version. The fifth paragraph on responsibility, authority, and communication, covers a part of 4.1 in the 1990 version. The sixth paragraph, on management review, is part of Section 4.1 of the 1990 version and includes input from a great many of the other sections as needed.

5.6.3 Section 6: Resource Management

The first paragraph on provision of resources is part of Section 4.1, management responsibility, of the 1990 version. Also, the resources question is mentioned in many other sections of the 1990 version. The second paragraph, on human resources, is mostly under Section 4.18, Training, in the 1990 version. The third paragraph, infrastructure, is assumed or mentioned peripherally under several sections of the 1990 version, such as 4.5, 4.7, 4.8, 4.9, 4.10, 4.11, 4.13, 4.15, and 4.16. The fourth paragraph, work environment, is also assumed or mentioned peripherally under several sections of the 1990 version, such as 4.5, 4.7, 4.8, 4.9, 4.10, 4.11, 4.13, 4.15, and 4.16.

5.6.4 Section 7: Product Realization

The first paragraph on planning of product realization contains part of Section 4.4, design control, and parts of Sections 4.6, 4.8, 4.9, 4.10, 4.12, 4.13, and 4.16 of the 1990 version. The second paragraph, customer-related processes, is partly new and also contains parts of 4.7, 4.15, and 4.19 of the 1990 version. The third paragraph, design and development, contains most of Section 4.4, design control, in the 1990 version. Purchasing, the fourth paragraph, covers 4.6, purchasing, and parts of 4.3, 4.10, 4.12, and 4.15 in the 1990 version. The fifth paragraph, production and service provision, covers Sections 4.7, 4.8, 4.9, 4.10, 4.11, 4.15, and 4.19, as well as relying upon 4.5, 4.12, and 4.16 in the 1990 version.

5.6.5 Section 8: Measurement, Analysis, and Improvement

Paragraph 8.1, in general, is an overview covering part or all of 4.10, 4.11, 4.12, and 4.20 of the 1990 version. The following paragraph, Monitoring and Measurement, includes checking up on the system as well as the product. It includes Section 4.17, specifically, plus parts of 4.10, 4.11, 4.12, and 4.20 of the 1990 version. New material on customer focus is included. The third paragraph, control of nonconforming product handles 4.14 in the old version by the same name and is on analysis of data. It is aimed at the new topics of customer satisfaction and continual improvement through data inputs from all sources about product and process. The fifth paragraph, improvement, is a new requirement. However, old Section 4.14 on corrective and preventive action has been included within Section 8.5. Previously, corrective and preventive action were considered to be emergency measures to handle process failures and prevent further failures. Note the change in philosophy. Now corrective and preventative action includes failure modes and effects analysis (FMEA), which is a forward-looking analysis to predict detrimental happenings on the basis of previous experience. Action on FMEAs is to be proactive.

Next is a summary of failure modes and effects analysis.

5.7 Failure Modes and Effects Analysis

5.7.1 Potential Risk-Avoidance Planning

- Characterize the part or the process.
- Ask how it might fail.
    - Learn from previous experience.
    - Perform a thought-experiment (brainstorming).
- List possible results of a failure.
- List the probabilities and risks of each possible result.
- List the deleterious consequences including costs of each risk outcome.
- List the potential approaches for corrective and preventive action.
- Make a decision on the approach to be instituted.
- Instruct the relevant organization to assign resources.

As an exercise for the reader, it is suggested that he/she try a failure mode and effects analysis on the example of the Cessna airplane given in Chapter 4, Section 4.2.6 (Point 6).

5.8 How Does NDT Fit into ISO-9000–2000?

There is much less emphasis on inspection in the new version of the quality management standard. It will be a matter of interpretation to justify the installation of inspection in the manufacturing plants of an organization under the new regime. The purpose of this book is to present the arguments needed to show that the use of inspection can be justified financially on a case-by-case basis. It is important for all NDT personnel, as well as all quality professionals, to understand the present quality management standard and to be able to work within it. The ISO-9000–2000 quality management standard is analyzed here section by section for the purpose of detecting where inspection, particularly NDT, fits.

The standard calls out for monitoring and measurement. Inspection is mentioned twice and testing is mentioned only once. One must draw inferences from the text as to where NDT or other tests would be acceptable if advantageous. The standard does not specify the exact kind of inspection.

Section 4 on the quality management system recognizes that the organization must identify the processes needed for its version of the quality management system. One can infer that the idea of inspection and NDT in particular should be brought into this thought process of identifying needed quality management processes. Then, if NDT or inspection have been identified as useful, the organization must ensure that funding and space are available for the inspection installation. When the manufacturing process is up and running in the factory, the organization must make measurements of the process on a continuing basis and analyze the data stream coming from the measurements. The analysis is both to check on the product and to check on the monitoring reliability.

Here, within the system itself, there are opportunities to insert NDT or other high-tech inspection tools, equipment, methods, and procedures. Everywhere the quality professional sees the injunction to identify the processes needed for the system, he or she should include the consideration of NDT. Whenever the quality professional is directed to ensure the availability of resources, he or she should not leave out NDT despite his or her training to cease reliance upon inspection. Where monitoring the process is called for, the idea of NDT to monitor intrinsic variables leading to latent flaws should come to mind.

Section 5, management responsibility, demands that management make a commitment to provide the funding for all the needed resources for all the processes assigned to it. This includes the measurement equipment and methods mentioned in Section 4 above. Management also has the responsibility to plan that the quality objectives are measurable. To do this, some of the output will probably need to be measured as identified above. To be measurable, measuring instruments and methods must be available. As the various processes do not go on smoothly forever by themselves, management is responsible for reviewing the system and finding chances for continuous improvement. For this management review, several pieces of input data are called for. When the review is complete, its output includes, among other things, a list of resources needed to tackle the situations encountered in the review. Management should not be surprised if some of these resources go toward inspection and testing.

It would appear that it is the responsibility of management to look into the utilization of NDT and to put it into effect if it appears to be advantageous. Again one sees the injunction to ensure the availability of resources, and NDT may be one among many areas needing resources. Planning should put the acquisition of such NDT resources up front so that the quality objectives, such as zero latent flaws, can be attained through measuring their otherwise undetectable presence. Top management should be thinking of ways to improve performance through NDT as it goes through reviews. Product conformity may often be enhanced through NDT monitoring, which may involve its use in preventative or corrective action.

Section 6, resource management, interacts with testing and inspection by demanding that management provide infrastructure to house and operate all processes. This implies that the management must provide buildings and workspace to house the inspection equipment it has found necessary in its earlier planning. It must also provide inspection equipment as well as process equipment if it has identified the need for such.

The quality professional in the role of team member in concurrent engineering will be expected to remind the committee to plan for the workspace to house the NDT systems as well as for the NDT instruments themselves to carry out the newly developed NDT inspections necessitated by the new manufacturing processes.

Section 7, product realization, indicates that management must plan for inspection and testing while planning the production of actual parts and final assemblies. This inspection and testing go along on an equal basis with other kinds of verification, validation, and monitoring activities to be performed on the product. In considering the customer, management should think of statutory and regulatory requirements. Some of these may be met best by inspection including, at times, NDT. Forward thinking in design and development of product should include planning for the verification and validation of quality at each stage of manufacture. NDT for verification-in-process could very well be a viable option in many cases. The development engineers should bear in mind statutory and regulatory requirements that may be best met by 100% high-tech inspection. Information from previous designs that may have needed testing will be invaluable. Lessons learned should not be forgotten. Reviews during development should consider all possible solutions to problems identified. The plans for inspection made in this section must actually be carried out for verification and validation of product. Inspection is mentioned explicitly as one method for verification of the quality of purchased product, which will go into processes in the organization. In the actual production operations, the organization must have measuring instruments available, must implement their use, and must carry out all release activities including testing, if planned, before shipments leave the factory. Special attention must be paid to processes that produce each part uniquely such that they are not amenable to SPC or ordinary verification processes. NDT is especially useful in these unique situations. As in all cases, the calibration and care of all measuring devices are critical.

Product realization (i.e., making the item) starts out with planning all aspects of the various processes including verification, validation, monitoring, inspection, and testing. Even if the consensus is that NDT is not required, it must enter into the thought process of the planners. Thinking further, some statutory and regulatory requirements might best be addressed through NDT to ensure safety of critical parts. As developments progress, product reviews should consider NDT in case it has been missed in the beginning or in case its utility becomes evident as developments go forward. Input raw materials or in-process inputs to further processes may need NDT attention. The NDT equipment and methods must be available for verification-in-process and for final release of product. In particular, NDT methods should be available for verifying product with respect to intrinsic variables where visual inspection or caliper measurements cannot detect the latent defects.

Section 8, Measurement, Analysis, and Improvement, calls for measurements to demonstrate conformity of the product to specifications and to suitability for use. Such measurements are to be carried out at appropriate points along the production line. Section 8 also addresses improvement. Within this topic, inspection may address problems in corrective action to find nonconformities and check on their possible recurrence. One might find that 100% inspection could address situations in preventative action, also.

Section 8 reiterates the need to be ready to use any and all measurement means to verify product and take corrective action if problems are detected. NDT methods are likely to be useful.

5.9 Summary

Inspection and inspection research and development (R&D) are valued in both the 1990 and 2000 versions of ISO-9000. Careful reading of both documents will show that there are many places where NDT measurements will be useful besides the places where inspection is mentioned explicitly. One hundred percent inspection of product by NDT will be the method of choice after analysis by certain FMEAs. This statement is made on the basis of experience. The need for verification-in-process to detect certain nonconformities may be met by NDT. Even verification of incoming raw materials and outgoing product may best be done, at times, by NDT. It must be remembered that ISO-9000 never specifies the methods, materials, machines, manpower, or environment to use in any industry or company. The standard only specifies that the product must be made well and kept fit for use. The quality management standard is replete with generalized instructions, but few are specific. NDT is never specified because it is a method that *may* be chosen rather than *must* be chosen. In the above paragraphs (at the end of the description of each section of the standard) listing places where NDT may be useful, this judgment is offered by the author and not taught explicitly by the ISO writers. It has been my experience that NDT has served expeditiously in many circumstances. Examples will be given in Chapters 7, 8, and 9.

6

Statistical Process Control as a Prerequisite to Calculating the Need for Inspection

6.1 Recapitulation of Statistical Process Control

Statistics in general and statistical process control (SPC) in particular are methods beloved by total quality management (TQM) and adaptable to ISO-9000, if not actually advocated (since ISO does not tell a company how to run its business). SPC lets the organization know when a process has gone out of control. In an after-the-fact fashion then, the organization learns information about the performance of the process while it was under control. In the most rigorous sense, the organization never knows that a process is under control. The organization is actually waiting for the process to go out of control at some unknown future time, which may be now, a few hours from now, or a few hours ago if the run rule, which will catch the out-of-control condition, takes several points (many hours) to operate to a conclusion. Over certain periods of time, known only in retrospect, the organization will be able to say that the process had been under control. The data amassed during those periods of time are critical to the calculations justifying or negating the use of inspection on 100% of production. One hundred percent inspection must be able to justify itself financially while the process is under control. In the special case in which a process is considered never to be under control, 100% inspection is mandated. Such a process turns out every part uniquely. No system can be devised to permit the definition of a good lot of parts, such as the group made in some other process monitored by SPC, before the SPC shows an out-of-control condition. Examples of this include in-mold inoculation of nodular iron (see Chapter 9) and forward extrusion of automotive axles.

The average fraction of parts that are nonconforming in the output of a process while it is under control is one of the three critical pieces of data to be used in decision calculations: to test or not to test. When the process is detected by SPC to be out of control, the process must be stopped, the parts made during the run rule must be quarantined, and those parts must be tested 100%. The decision about testing *all* parts *all* the time depends upon the proportion nonconforming parts made while the process is under control. This proportion is found in retrospect over extended periods of time while the process was in control.

Hence, in this book the use of SPC is advocated on a continuous basis as a precursor to a decision to do 100% inspection of all of production. SPC should still be used while the 100% inspection is going on. Stopping the process for repair, and then quarantining and testing the product made during out-of-control conditions, are still necessary actions if the decision is not to test. If 100% testing is going on, and if the data on each part are recorded (which often is not done), then each nonconforming part could be culled from production on the basis of the 100% test. SPC would still indicate when to stop the process. It is possible that the output of the 100% inspection might be used in the SPC formulas to find SPC data points periodically (Papadakis, 1990). However, simply finding some nonconforming parts by inspection is not proof of an out-of-control condition. The SPC process from Chapter 3 must be used on the inspection data if it is to be used for SPC. It is advocated that SPC be used continuously while financially justified 100% inspection is also used. Certain pieces of data derived from the process while it is known to be under control will be used to continuously check whether the 100% inspection is still necessary.

6.2 Necessary Data

6.2.1 Rate of Production of Nonconforming Parts

As mentioned above, the rate of production of nonconforming parts is the basis of the calculations for financial justification of 100% inspection. One needs to know the proportion of nonconforming parts (p) produced on the average over time. A proportion is a fraction like 2/10,000 or 1/25, and may be expressed as a decimal for purposes of calculation, like 0.0002 or 0.04. It is empirical, measured over a long time while the process was under control. It may be the average proportion of nonconforming parts over several shorter periods while the process was under control, of course. In other forms of calculation, one may need integrated figures like 1000 nonconforming parts per year or such. It is possible that data or projections over several years may be needed in cases involving investments in equipment to be amortized over time. The details will be given in Chapters 7 and 9.

6.2.2 Detrimental Costs of Nonconformities

The detrimental cost (k_2) to the organization of one nonconforming part's going further into production is the second datum needed for the financial calculation. Elements contributing to this cost are as follows. First, consider that if a nonconforming part is not detected, it will proceed further into production and generate costs for processing it further. At that point it will be mixed with conforming parts and possibly some other nonconforming parts that have undergone further processing. At that point, if one were to discover that nonconformities had slipped through, there would be a second type of cost—the cost of sorting this entire lot to pick out the nonconforming parts so that more processing would not occur and be wasted. If the parts went further into an assembly before detection of the existence of nonconformities, then there would be a third kind of cost—the cost of disassembling the assemblies, repairing them with good parts, and reassembling them. (A subsidiary but not insubstantial cost might have been incurred to ensure that the repair parts were not nonconforming.) A fourth kind of cost would be incurred if the repairing of the assemblies took so long that production of larger assemblies scheduled to use the assemblies now under repair had to be delayed. (I experienced one of these events in an automobile factory. Twenty thousand transmissions required repair and rebuilding, shutting down an automobile assembly line and costing the company $5,000 in profit for each car delayed at a scheduled production rate of 60 vehicles per hour. Repairing the transmissions required many hours.) Even worse, if the parts got out into salable product and were detected only in the field during customer operation, the fifth kind of cost—warranty costs—would take effect. Many types of failures during customer operation require recalls, and a sixth kind of cost is the cost of those recalls where an inordinately large number of devices, say vehicles, must be located, their owners notified, and the parts replaced at the manufacturer's expense. If the failure of the part caused equipment outages, then a seventh type of cost is incurred—the cost of repairing the outage compounded by lost production during the repair process. An example of this was the PGM tube bursting in the paint bath (Papadakis, 2000a) cited in Chapter 4, Section 4.2.6. Legal situations involving alleged damage to plaintiffs (customers or third parties) may arise, and lawsuits may yield an eighth type of cost. Totally elusive from a quantitative standpoint, but very detrimental, is the ninth type of cost—the loss of reputation due to negative comments about your product by dissatisfied customers. Any actual loss involving a customer can probably be doubled when taking this phenomenon into consideration. One or several of these costs may be operative in any given case.

Sometimes the calculation will use an integrated value like the total detrimental cost in a year, for instance. For investment methods, one will need costs and projections for more than one year. One can see that the detrimental cost can escalate depending on how far the part goes beyond its point of manufacture, and on how critical the part is in the ensuing structure. To keep nonconforming parts from going too far, the inspection along the production line would be termed verification-in-process (VIP).

The worst-case scenario would involve a part, the failure of which could bring down an airliner or sink a submarine. Such things have happened in cases where NDT during production or servicing could have detected the nonconforming part. One case was the failure of a turbine disc in a jet engine in United Airlines flight 232, a Douglas DC-10 that crash-landed in Iowa in 1989 with more than 50% loss of life. The airliner lost power to its control surfaces because the disc, breaking at more than 20,000 rpm, sliced through the hydraulic tubing in the vicinity of the engine. The disc broke because of a crack. It was not clear at the time whether the state of the industry in NDT of aircraft engines would have found the crack before it grew to criticality. Another case was the loss of the USS *Thresher* in the Atlantic in 1963. Running under the surface, the submarine was flooded by a series of events initiated by the failure of a poor braze on a pipe handling seawater in the engine room. A colleague of mine from Automation Industries (Bobbin, 1974) had proved shortly before that an ultrasonic test, adopted but not systematically used in the Navy at the time, could have detected the bad braze (EH9406, 1994). While these cases may seem like rather insignificant statistics compared with mass production, there are thousands of engine discs made per year and thousands of marine welds and brazes, too. Failures surely occur in automobiles where millions are manufactured annually, but hardly any failures are ever diagnosed down to the metallurgical or mechanical failures amenable to production inspection. I suggested (Papadakis, 1976a) that a mechano-coroner be attached to every county court to act in mechanical accidents as a coroner acts in human fatalities. Not known to the public are the great efforts in inspection motivated by failure modes and effects analysis (FMEA). The engineer developing a financial calculation to justify 100% NDT in production should study the above information and all the possibilities within his industry.

6.2.3 Costs of Inspection

The cost to test one part, k_1, is the third datum needed for the financial calculation. Elements contributing to this cost are as follows.

First, one must consider capital equipment. Several costs come under this heading. There is an *initial cost* to purchase the equipment. If this is a large amount and is to be amortized over several years (depending on the tax code), there is *depreciation* to consider. If an endpoint of the utility of the equipment is projected, then there is *residual value* to consider. An endpoint will be predicted by the *actual cycle life* of the design of the part to be tested. (For instance, an engine may be phased out after 3 years, so the production line where the test equipment is installed would be shut down.) *Planned production volumes* must be addressed to determine how many pieces of NDT equipment might be needed. The *cost of capital* must be factored in because the decisions about capital purchases are made on that basis.

Second, one must consider operating costs. Among those are *labor*—the grade or level of the needed equipment operator or machine tender must be considered. Because the test station will take up space, equivalent *rent* must be calculated. Utilities attached to the equipment and used during the year, like kilowatts, must be accounted for. Some *maintenance* will be needed, which may be done in-house or on a service contract.

Third, there is the possibility of subcontracting the job to an outside service company. That company's *piece cost* would have to compete with the comparable cost obtainable in-house.

Sometimes the calculation will use an integrated value like the total testing cost per year, for instance. For investment methods, one will need costs and projections for more than one year. The engineer developing a financial calculation to justify 100% NDT in production should study the above information and all the possibilities within his industry.

6.2.4 Time until Improvement Lowers Nonconformities

One cost–benefit principle used in 100% inspection is that the inspection must pay for itself and save money in the period of time during which it is still needed. For a short-run need, the organization might opt for a manual operation with cheap instruments rather than choosing to invest in expensive automation with high-end electronics.

The run length can be determined by many things. One would be the length of time during which the part will still be produced. If a part of one material were to be superceded in 6 months by a part of another material, then testing the first material would not justify a long-term investment. Change of models could be as important as change of materials.

Continuous improvement provides a more involved calculation, estimate, or possibly negotiation. TQM and statistical people may believe that continuous improvement will obviate the need for inspection in a short period of time, say 6 months. The savvy process engineer might estimate 1 year at least. The inspection technologist, having seen cases like this drag on for years, might hold out for 2 years but really believe 3 years, having seen how slowly the organization's research arm operates. In a team doing concurrent engineering for continuous improvement, the committee chair might be conservative and be willing to invest in inspection for 2 years. Assuming that 100% inspection could pay for itself at all, would the inspection pay for itself in that time with automation or would a manual work station be used? Judgment may supersede rigid formulas or mantras.

If the TQM personnel prevailed and then continuous improvement failed, who would pick up the pieces? In the case cited in a previous chapter (Chapter 4, Section 4.2.9) where the vendor company promised perfection through continuous improvement and the organization bought off on the vendor's assertion, the organization's in-house inspection technologist was ready to pick up the pieces by having a newly developed test ready to go because of timely concurrent engineering started 2 years before Job 1.

Chapter 9 will give many examples of 100% inspection on the production line where manufacturing improvements made over time brought the proportion of defective parts (p) down, but not down far enough to permit the elimination of 100% inspection. The idea of continuous improvement should be taken with a pillar of salt for fear that the product will be left behind like Lot's wife.

6.3 The Costs of Inspection and the Detrimental Costs of Not Inspecting

The costs to inspect parts are listed in Section 6.2.3. One specious cost the manufacturing people attempt to charge against the inspection technologists is the cost of throwing away faulty parts. The manufacturers want to ship everything. If the transfer price of a part is $10, then the manufacturing engineers will attempt to charge $10 to the inspection department for every nonconforming part detected and rejected. The company should simply absorb the cost. The reality, of course, is that rejecting the $10 part probably saved the organization from $100, $10,000, or even more in warranty, lawsuits, and damaged reputation (these detrimental costs are covered in Section 6.2.2). By contrast, the cost of testing in order to reject the part was probably on the order of $0.10 per part.

The damage to company reputation is impossible to quantify. Dr. W. E. Deming emphasized the critical importance of the loss of reputation because of poor quality. Studies have shown that detrimental experiences are mentioned by customers much more frequently than are pleasant experiences. I have gotten lots of mileage at cocktail parties telling about a crew of ace mechanics who would not believe, until my third return to the garage, that I had melted down the ceramic liner of an automotive catalytic converter when an engine control computer failed on a four-cylinder engine, letting it work like a one-cylinder lawn mower engine sending raw fuel–air mixture through the hot exhaust system. I have also mentioned innumerable times my success at getting a refund for five clutches and a flywheel, after a different ace mechanic finally determined that the abnormal wear had been from a manufacturing defect. Of course these examples have nothing to do with inspection, but they illustrate the principle of reputation. Ruining one's corporate reputation can be a cost of not inspecting. Not applying optical shearography to tires may have damaged some reputations in connection with recent SUV (sport utility vehicle) rollovers. I learned from a shearography salesman that the Israeli army was getting 20,000 extra miles out of truck tires by such inspections. This is in maintenance, not manufacturing, but the example is interesting. In the case of SUV rollovers, shearography might have been useful in the manufacturing of tires.

The types of high-tech, 100% inspections related in this book are not found in either the TQM literature or in the ISO-9000 standards. The TQM quality professionals adhere to the Deming points about not relying upon inspection and about doing continuous improvement, with SPC methodology interspersed throughout. The ISO standards writers do not tell a company how to run its business. They would no more specify NDT equipment than insist upon electric furnaces or hydraulic presses. The work with NDT equipment and its incorporation into 100% inspection as expressed in this book is complementary to the TQM and ISO philosophies. The 100% inspection philosophy incorporating NDT and financial calculations is a product of my professional experience and expertise. The financial calculations in Chapters 7 and 9 will stand by themselves as evidence for the utility of 100% inspection in manufacturing.

6.4 Summary

SPC should be carried out on processes before the financial calculation is done with respect to the need for 100% inspection. SPC will indicate when the process was under control, after the fact. From the data taken while the process was under control, the correct value of p, the proportion defective, will be found and incorporated into the financial calculations. The costs for k_1 and k_2 will have to be found from experts or in the company archives.

7

Three Financial Calculations Justifying 100% Nondestructive Testing

7.1 Introduction

There are three principal financial methods for calculating the propriety of choosing to perform 100% inspection on an item of production. In each method, the answer may come out yes or no. Before going further, it must be stated that the inspection method itself must be nondestructive. Otherwise, one must revert to batch certification by statistical methods applied to destructive tests, that is, to sampling. Sampling is well known and will not be dealt with in this book. Nondestructive testing (NDT) methods and correlations will be reviewed in Chapter 8.

The key to each financial method of calculation is that the detrimental costs of not testing outweigh the costs of testing. The outlay for inspection is expected to terminate when continuous improvement has lowered the overall detrimental costs to a point where they do not exceed the costs of testing. This may never happen of course, although hope springs eternal that it will. The financial calculations must include the assumption that the investment in the inspection equipment will pay for itself in the period of time before adequate improvements are completed and before the production of the part is terminated.

The three financial methods appear below. The titles are presented succinctly in Table 7.1.

7.1.1 The Deming Inspection Criterion (DIC) Method

This method uses the cost of inspecting each part, the detrimental cost if one nonconforming part goes further into production, and the fraction of nonconformities known from experience to arise from the production process to determine when to do 100% inspection. This method is best for inspection technologies where the equipment investments can be written off in one year and where the major expense is variable costs. It is also useful where the

TABLE 7.1

Numerical Methods for Justifying 100% NDT

(1) BREAK-EVEN: The Deming Inspection Criterion
(2) INVESTMENT: The Internal Rate of Return or Time-Adjusted Rate of Return
(3) PRODUCTIVITY: Productivity, Profitability, and Revenue (Quality, Productivity, and Profit leading to improved competitive position)

inspection is done by a vendor who will quote piece costs. Integrated values of the cost to test for a year and detrimental costs accrued for a year may be used along with the proportion defective.

7.1.2 The Time-Adjusted Rate of Return (TARR) or the Internal Rate of Return (IRR) Method

This method is good for the situation in which the investment in the inspection equipment and its automation is large and will be written off over several years of use during which there will also be variable costs. The data include the rate of production of parts, the rate of production of nonconformities, the detrimental cost per nonconformity going further into production, the lifetime of the inspection before it is rendered unnecessary by continuous improvement, the residual value of the equipment after that time, and the interest rate the organization is willing to pay on money borrowed to purchase capital equipment.

7.1.3 The Productivity, Profitability, and Revenue Method

This method traces dollars earned vs. dollars expended by any process in terms of productivity written as dollars per dollar in an input–output equation where all resources are translated into currency equivalents. The detrimental costs of nonconforming products going further into production reduce the dollars earned (numerator) and hence reduce productivity. Inspection reduces the total detrimental cost while increasing production costs.

The net calculation can increase productivity and profitability, resulting in increased revenue.

The calculation algorithms will be presented in this chapter and examples will be given in Chapter 9.

7.2 DIC: Low Investment

The equation for the Deming inspection criterion (DIC) is

$$\text{DIC} = (k_2/k_1) \times p \tag{7.1}$$

where k_2 is the detrimental cost of one nonconforming part going further into production, k_1 is the cost to inspect one part, and p is the proportion (fraction) of production that is nonconforming. Various potential sources of the detrimental costs k_2 were written down in Section 6.2.2 while components of the inspection costs k_1 were listed in Section 6.2.3. The reader is referred back to Chapter 6 to study these costs.

Equation (7.1) is the solution to a problem for the student in a classic quality treatise (Deming, 1982) and was solved in the text of the advanced for-revision versions of that book (Deming, 1981) used previously in Deming's four-day course on quality management (Walton, 1986). Several examples of its use in proving the necessity of 100% inspection were given by E. P. Papadakis (1985a). Deming gives other examples in Chapter 13 of his 1982 book. In the paper by Papadakis (1985a), continuous improvement was shown to be inadequate in some cases to negate the need for 100% inspection despite long periods of application of engineering for improvement.

In order to use Equation (7.1), the process producing the item in question must be under control. The use of SPC (statistical process control) is advocated in Chapters 3 and 6 for ensuring that the process is, indeed, in control. If the process is out of control, Equation (7.1) may still be used if it can be determined that the process is intrinsically never under control or that the time to gain control of the process will be long in terms of the continuing production of nonconforming material. The concept of a process intrinsically never under control was addressed in Chapters 3 and 6. The time scale is measured, also, by the installation and operation of the inspection method. In other words, the time must be long enough for the inspection effort to do some good. As stated previously, the current concept is that the material produced while a process is out of control must be inspected to eliminate nonconforming material.

When the data are inserted into Equation (7.1), the inspection decisions are as follows:

$$\begin{aligned} &\text{Yes} \quad \text{for} \quad \text{DIC} \geq 1.0 \\ &\text{No} \quad \text{for} \quad \text{DIC} < 1.0 \end{aligned} \tag{7.2}$$

The higher the cost ratio k_2/k_1 is in Equation (7.1), the lower the proportion of nonconforming p must be to preclude the need for 100% inspection. For instance, if $k_2 = \$10{,}000$ and $k_1 = \$1.00$, then $k_2/k_1 = 10{,}000$ and p must be less than $1/10{,}000$ for no testing. If p is greater than $1/10{,}000$ (0.0001), then testing is called for. Further examples will be given in Chapter 9 for real production cases.

7.3 TARR or IRR: High Investment and Long-Term Usage

These methods calculate the interest rate to be realized on an investment to be made at time zero and used for several years. Every company controller is fully familiar with these methods and has canned software to perform the calculations if given the data. The method makes a comparison between an existing situation and a new situation brought about by the investment. The method can be used on any new investment, such as a new factory to replace old facilities, a super tanker to replace four Liberty Ships, a new heat-treating furnace to replace an old one, a machine to replace manual operations, or inspection apparatus to replace warranty expenditures. The principle is that if the current practice is continued, one stream of costs will accrue year by year; if a new practice is instituted, a different stream of costs will accrue. The different stream is the result of the investment item put in place at time zero. After the streams are projected out a certain number of years, the two streams can be used as data in the IRR program to determine if a net savings would result, and to determine what effective rate of return would be earned on the investment. This method was formally introduced into the inspection and nondestructive testing business by Papadakis et al. (1988).

In the case of investment in inspection equipment, for instance involving an NDT instrument with associated automation, the operating costs yearly are an expense and the income tax savings due to depreciation are on the positive side. This stream would typically be compared with warranty costs if the inspection equipment were not installed to eliminate the nonconforming material with real or latent defects that might fail. Other detrimental costs from Section 6.2.2 could accrue. Two typical cost streams to be compared are shown diagrammatically in Figure 7.1. With real numbers inserted,

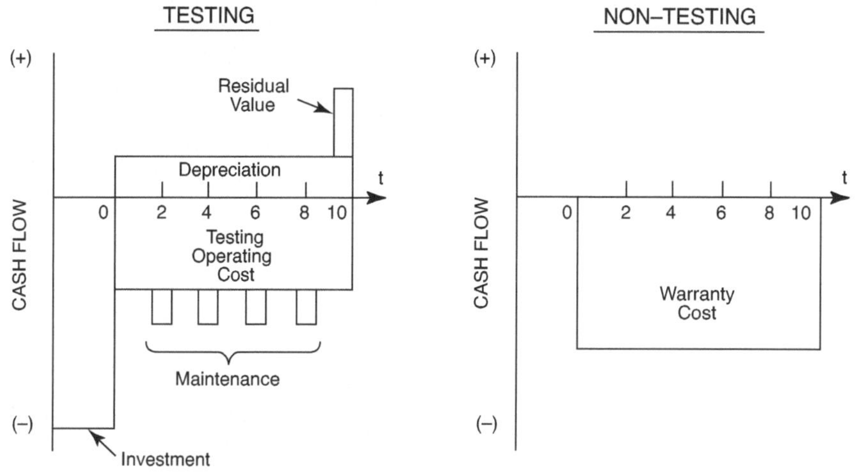

FIGURE 7.1
Two cost streams to be compared by the method of Time-Adjusted Rate of Return or Internal Rate of Return to determine whether to purchase inspection equipment for use over several years. (Reprinted from Papadakis, E. P., Stephan, C. H., McGinty, M. T., and Wall, W. B. (1988). "Inspection Decision Theory: Deming Inspection Criterion and Time-Adjusted Rate-of-Return Compared," *Engineering Costs and Production Economics*, Vol. 13, 111–124. With permission from Elsevier.)

the factory controller could calculate the interest to be earned by investing in the inspection equipment. He could then decide if the investment was feasible by comparing the interest rate with the *hurdle rate* specified by the company. This is a variable figure depending on the overall economy and the financial health of the company.

7.4 Productivity, Profitability, and Revenue Method: Nano-Economics

I pioneered this method in the mid-1990s (Papadakis, 1996). The method is a quantitative expression in four equations of the title of Deming's 1982 landmark treatise, *Quality, Productivity, and Competitive Position*. The thesis of this book can be stated as a three-line promise, as follows:

> If you increase quality,
> You will raise productivity, and
> Improve your competitive position.

For the equations, the three lines are expanded as follows:

> If you increase quality by lowering nonconformity proportion,
> You will raise productivity, and
> Get more revenue to spend on any appropriate strategy to improve your competitive position.

The actual equations are given here:

$$P = (A - B)/C \tag{7.3}$$

$$E = P - 1.0 \tag{7.4}$$

$$D = E \times C \tag{7.5}$$

$$G = \Sigma D \tag{7.6}$$

The first three equations refer to any single process within a factory, while Equation (7.6) is the sum over all the processes in the factory. The equations must be understood in terms of the two diagrams of a process shown in Figures 3.1 and 7.2. Figure 3.1 shows the main branches of the wishbone diagram of a process working inside a boundary and producing an output. From here the next critical step is to understand from Figure 7.2 (Papadakis, 1992) that the process uses up resources as inputs labeled as C = Value In, while having two outputs, A = Value Out and B = Disvalue Out. The quantity A is the value for which you can sell the output, namely the number of pieces N times the transfer price T, or

$$A = N \times T \tag{7.7}$$

On the other hand, quantity B is the sum of all the detrimental costs that come about because of the production of n nonconforming parts among the N. The causes of the detrimental costs are again from Chapter 6, Section 6.2.2. Calling V the detrimental cost per part, then

$$B = n \times V \tag{7.8}$$

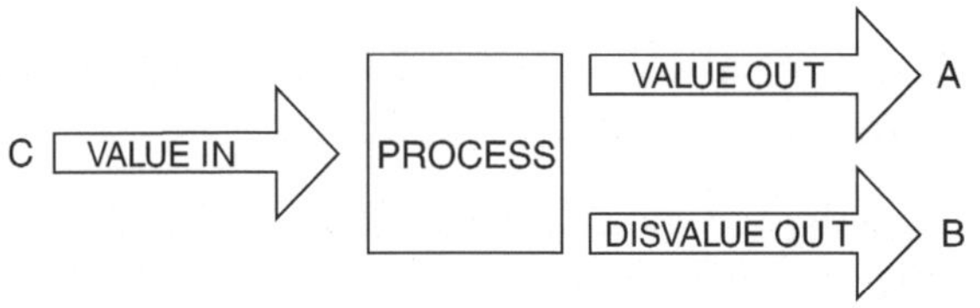

FIGURE 7.2
Diagram of value flow through a process. The value C-in runs the process. The value A-out is the revenue from the sale of its output. The disvalue B-out is the detrimental cost of having nonconformities in the output. B can become very large if the potential cost of a single nonconformity is large. (Copyright 1992 © The American Society for Nondestructive Testing, Inc. Reprinted with permission from Papadakis, E.P. (1992). "Inspection Decisions Based on the Costs Averted," *Materials Evaluation*, 50(6) 774–776)

The quantity V is somewhat like B. Hoadley's (1986) value-added-detractor (VADOR).

The value E is the economic profitability of the process, and is 1.0 less than the productivity P. If productivity falls below 1.0, then the process begins to lose money.

The dollars D are realized from the process as profit and are calculated as the economic profitability E times the cost C to run the process. This amount becomes negative if the profitability E becomes negative.

Finally, the gross profit G for the factory is the sum of the values of D for every process.

The consequences of poor quality can be analyzed as follows. Since the detrimental costs associated with poor-quality items can be very high, it is possible to have $V \gg T$ while also having $n \ll N$ (i.e., even for high capability). This pair of inequalities indicates that the value of B can be comparable to the value of A. Further, this means that productivity P could go to zero or even become negative. Economic profitability E could be zero or negative, and the revenue D could become negative if V were large enough even for high process capability (n is very small in percentage).

Inspection fits into this regimen by being capable of making B "out the back door" essentially zero. NDT fits into inspection because many latent defects can be detected only by NDT methodologies. Modern high-tech inspection methods such as NDT can also accomplish a much larger reduction in B than could human inspectors with visual inspection, calipers, and so forth. The common wisdom is that inspectors were only about 80% effective in their sphere of operation. In addition, high-tech methods can detect and measure intrinsic properties of matter as well as the extrinsic properties the human inspector could sense, enabling a much broader improvement in quality with the employment of high-tech instruments.

Inspection will add some cost to the production costs C and will lower the number of salable items from N to $N - n$, reducing the value A. Extra production, possibly at overtime rates, will be needed to fill the contracts for N items. Thus, while using inspection, the productivity will be somewhat lower than for perfect production, but certainly higher than if B were allowed to remain large.

In Chapter 9, examples of striking improvements in productivity and profitability due to inspection will be shown despite long efforts at continuous improvement. It will be obvious that inspection should be instituted and continued in certain calculable cases.

In summary, the three methods for calculating the propriety of using 100% inspection have been outlined and analyzed. They lead to unambiguous and unbiased objective results and can be used as proof in the presence of differing opinions.

8

High-Tech Inspection Methods

8.1 General

8.1.1 Documentation and Methods

The very notion of a high-technology inspection method is beyond the scope of the originators and adherents of total quality management (TQM) and of the control-chart advocates who brought statistical process control (SPC) into being. This statement is evidenced by the absence of a category for nondestructive testing (NDT) and all its synonyms in the index of Dr. W. E. Deming's magnum opus (1982) and the Western Electric Co. handbook (1956). W. A. Shewhart's book (1931) was written before formal NDT. Point 3 of Deming's Fourteen Points of management (Table 4.1) advocates the elimination of dependence upon mass inspection without acknowledging that ongoing inspection is useful at times. To his credit, Deming (1982) does mention three basic circumstances where inspection should be performed. These are (1) parts critical for safety, (2) new or changed parts (including new production venues) where testing should go on for 6 months to obtain data, and (3) parts where cost analysis based on variants of the Deming inspection criterion (DIC; see Chapter 7 in this book) show that money can be saved. This third item occupies Chapter 13 in Deming's 1982 text (pages 267–311). However, as late as 1984, Deming stated that he did not know of the capability of instruments based upon physics and electronics to detect latent flaws (Deming, 1984). He also stated that he did not know that high-tech instruments could measure intrinsic variables for constitutive equations that would predict future behavior (failure) of materials. In the case of the Western Electric handbook, the authors stick to the subject of SPC assiduously, whereas it is well known that other people within AT&T, the Bell Telephone Laboratories, and Western Electric knew of and practiced NDT. See, for instance, the classic book by W. P. Mason of Bell Telephone Laboratories, *Physical Acoustics and the Properties of Solids* (1958). On the very first page he acknowledged NDT with the statement: "If… imperfections are present, they cause reflections or refractions of the sound pulses. These reflected or refracted waves produce responses, arriving after the time of the sending pulse, which can be picked up by the same or an adjacent transducer. Hence these pulses provide a means for examining or inspecting the imperfections of a solid body. Ultrasonic inspectoscopes and thickness gauges are among the best devices for determining the integrity and dimensions of metal castings or other solid bodies."

This lack of modernization of quality concepts among quality professionals went on unabated even as the testing and inspection professionals continued to publish papers on new methods and to update their handbooks for engineers and technicians. It should be noted that the two volumes of the first edition of the *Nondestructive Testing Handbook*, edited by R. C. McMasters for the American Society for Nondestructive Testing (ASNT; 1959), were tours de force in collecting and explaining both useful methods and theory as far as it had been developed to that date. Much of the text is still valid, and the book is still in print.

The process of collecting and explaining continued. In 1982, the ASNT received the copyright on the first volume of the second edition of its *Handbook of Nondestructive Testing*. This edition came out in ten volumes, the last one in 1996. These volumes are edited by P. McIntire and others. As time progressed, a third edition of the ASNT *Handbook of Nondestructive Testing* was initiated. The first volume of the third edition bears a 1998 copyright. Other volumes have been issued, and still others are in preparation. The editor finds the most experienced practitioners in the subfields to write the volumes. All volumes mentioned above (including 1959) are still in print and are available through the ASNT publications catalog (ASNT, 2005).

Concurrently, other organizations issue handbooks and compendiums on nondestructive testing. The American Society for Testing and Materials (ASTM) issues an updated version of Volume 3.3, "Nondestructive Testing," of its set of books on specifications and recommended practices annually. New documents for inclusion and revisions as needed are being voted upon continuously. The American Society for Metals (ASM) updates its *Metals Handbook* on a long-term basis. One volume (ASM, 1976), Number 11 in the eighth edition, is entitled *Nondestructive Testing and Quality Control*. In addition, many individual authors publish books on specialized topics.

The concept of using an NDT instrument to find intrinsic variables and latent flaws is shown diagrammatically in Figure 8.1. At the same time, SPC statisticians (of whom Deming was one) did not propose the use of anything but simple analog instruments (rulers, scales, and fixtures), which could only measure extrinsic variables such as length and weight. SPC and TQM were ostensibly satisfied with poking around the tip of the iceberg of quality, so to speak, not knowing anything hidden below the waterline. This meant that 90% of quality was off limits to the salutary activity of SPC and TQM. Of course, quality was so bad in American manufacturing that the quality professionals had their hands full and would have made great progress simply by getting extrinsic variables under control. That was the state of affairs to which the systematic application of Frederick Taylor's philosophy had brought manufacturing in the United States.

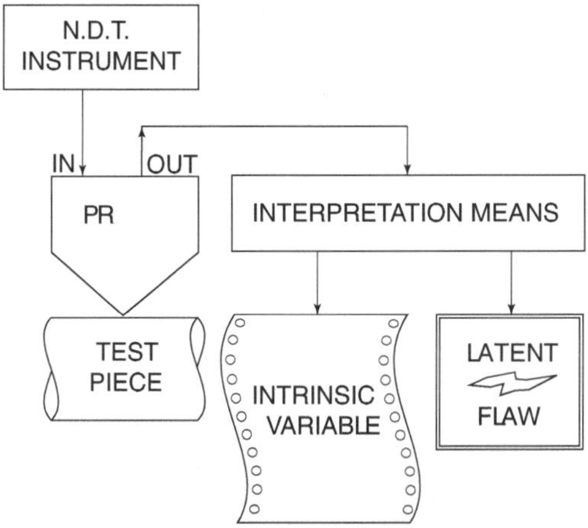

FIGURE 8.1
The concept of using an NDT instrument to find intrinsic variables and latent flaws.

A very interesting phenomenon occurred concurrently with the 1930s approach of TQM and SPC. Physicists and all sorts of scientists using applied physics attacked the question of probing for latent flaws and intrinsic variables. They wanted to find a latent flaw without destroying a part. They wanted to predict the intrinsic variable associated with a piece of material without needing to fabricate and break test pieces such as tensile bars, Charpy bars, Izod bars, spring-back sheets, and so on. Applied physics provided the methods to accomplish these ends in many cases. The ideas and methods evolved seamlessly into NDT over a period of time. The applied physics was turned into an industry (NDT) by entrepreneurs who sold solutions to problems, not just instruments and devices. The process of developing pure science into salable NDT solutions is shown in Figure 8.2. For instance, instrumented systems could be as large as self-propelled railroad cars that tested railroad track *in situ* on a service contract basis in the 1940s and 1950s, and still do (See Chapter 10, Section 10.3).

A very early example involves piezoelectric crystals. When piezoelectricity was discovered, it was shown to be reciprocal. That is, a body would compress or expand if an electric field was applied through it. If stress were applied to the body, a voltage would appear on some of its faces. Crystals having this property were fabricated into devices to transmit and receive sound waves, which are stress waves, of course. One of the first NDT applications of this device was sonar in the First World War. The French were performing NDT of the oceans for unwanted inclusions or flaws, namely German submarines. As the crystals used at the time were water soluble, the transduction devices had to be encapsulated. The field has developed such that all NDT ultrasonic transducers consist of encapsulated crystals or ceramics except for a few, which are electromagnetic coils.

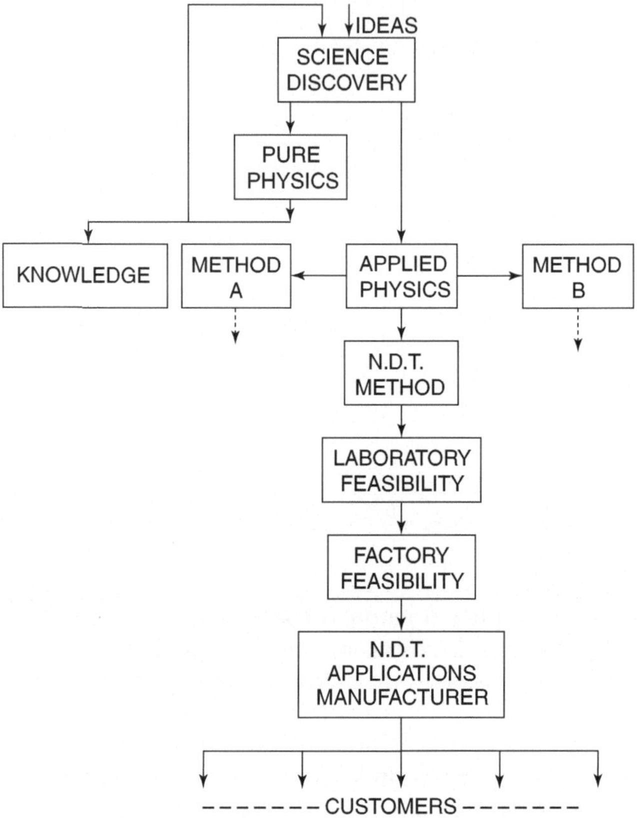

FIGURE 8.2
The process of developing pure science into salable NDT solutions.

Another very early example was the discovery of Roentgen rays (x-rays). Radium was used at first; soon, high-voltage vacuum tubes were invented to produce x-rays. Medical diagnostic applications on bone fractures and NDT applications to find cracks and voids in inanimate objects developed in parallel quickly. This is an example of Methods A, B, and NDT coming out of applied physics in Figure 8.2. The American Radium and X-ray Society was formed in 1940 to expedite the applications to metals, in particular. Tank armor and such things were tested regularly. As other methodologies were incorporated into testing, the society changed its name to the American Society for Nondestructive Testing.

Another method that came to the fore early was eddy currents. Eddy currents were discovered almost as soon as transformers for alternating current. While transformers use a magnetically soft iron for their core between two coils, eddy current instruments use any piece of metal to be tested as if it were the core of the transformer between two coils. Some electrical and magnetic properties of the piece of metal can be deduced, and cracks near its surface can be detected. Eddy current methods became an integral part of NDT.

Surface-breaking cracks can be detected by dye penetrants, which show the cracks as colored lines, and by magnetic particles, which are held on the cracks by magnetic fields jumping from one side of a crack to the other. The magnetic particles usually carry a dye that fluoresces under ultraviolet light for visibility.

These five methods, x-ray imaging, ultrasound, eddy current, dye penetrant, and magnetic particle, became known as the Big Five of NDT by about 1960. See the summary in E. P. Papadakis, 1980. In actuality, the methods did not all become accepted simultaneously, but rather there was a phenomenon of entrenchment and breaking in, so to speak. As a method became accepted and standards were written around it (ASTM, 2005), another method had to prove its worth by vigorous endorsement by advocates as well as by rigorous testing. For instance, in 1958 the use of ultrasonic inspection to find flaws inside bodies was having a difficult time being accepted because of the preference of inspection personnel for tried-and-true x-rays (McEleney, 1958). That year, personnel at the Watertown Arsenal developed a pioneering multimodal method using ultrasound and eddy currents to test for improper heat treatment of gun barrels. Many more methods are now available besides the Big Five. The other fields and methods are growing so that in the not-too-distant future there may be a Big 7 or a Big 11; change is the only constant. The ASNT recognizes this fact in issuing new subject volumes in its *Nondestructive Testing Handbook* as methods are developed. In particular, acoustic emission (AE) as a method has always been treated as separate from ultrasonic testing because AE arose and matured later and was passive instead of active with respect to radiant mechanical energy in the ultrasonic range. Several methods will be explained in detail in this chapter for the benefit of quality personnel coming to the financial calculations in Chapters 7 and 9 from a background not strong in NDT.

Here it is emphasized that the NDT methods were adopted by many industries including automotive, defense, and aerospace essentially as soon as the NDT technology was shown to have factory feasibility as well as technical feasibility. See the explanation of these terms in Section 8.1.2; examples will be given. NDT became necessary for the production of parts as well as for the safety of the people using the parts. The ASTM standards book (Volume 3.3 in their series, ASTM, 2005) is a compilation of recipes for testing items. The ASNT *Nondestructive Testing Handbook* has more science content (ASNT, 1959).

As TQM and SPC began to move into the quality sphere in the 1980s, the TQM and SPC personnel had not caught up with the technology of NDT and some other high-tech inspection methods. The Deming dictum about eliminating dependence upon mass inspection came just as NDT was blooming into the method of choice to find latent defects (Papadakis, 1980). The Deming dictum led to the dismantling of much useful inspection technology and to the failure to install even more. This dictum was pursued without the knowledge of the benefits of NDT (Deming, 1984). For instance, the Ford Motor Company disbanded its NDT research and development group in a major reorganization in 1985 after adopting the Deming philosophy. It became obvious that continued advocacy of NDT would have detrimental career consequences. At General Motors (GM), one product manager (Bloss, 1985) determined to keep his NDT operations viable despite TQM by using NDT within the concept of verification-in-process in the upcoming ISO regimen.

To fix ideas without becoming too technical in this section of the chapter, it should be mentioned that the two entire issues of *Materials Evaluation* (*ME*) for November and December 1984, were dedicated to automated NDT. There were 15 refereed, abstracted, and archival technical papers in the two issues. Various major manufacturing industries supplied papers. This author was the guest editor for the special topic. *ME* is the technical journal of the American Society for Nondestructive Testing.

Not all NDT is automated, of course. Some is manual. Using the results in Chapters 7 and 9, one can choose the more practical scenario: automation or manual applications.

8.1.2 Definition and Outlook

NDT is defined loosely as all the methods of testing an object to ensure that it is fit for service without damaging it and making it unfit for use. The presumption is that certain classes of mechanisms that would make an object unfit for service can be detected by nondestructive applications of physics embodied in electronic devices.

NDT is an amalgam of three inseparable aspects: methods, instruments, and intelligence. Methods are developed by intelligent people using theory as a guide and employing instruments for experiments. Tests based on these methods are developed and embodied in instruments for use in practical situations in two steps:

1. *Technical feasibility,* which shows that the method could yield desirable results in a laboratory on good parts vs. parts known to have the nonconformities the engineer desires to eliminate.
2. *Plant (factory) feasibility,* which shows that the feasible laboratory test is robust in the sense that it could be used in a harsh environment and could still detect nonconformities unambiguously in the presence of all the variables in a factory. Note the Four Ms and the environment in Figure 3.1, each of which provides variables in a plant (factory) situation.

Note the sequence in Figure 8.2. Then tests based on the methods are developed for specific environments (customers) and are carried out either by people using instruments or by automated systems. Intelligence is required for the interpretation of the output of instruments. The intelligence may be supplied directly by a certified operator (ASNT, 1988) or indirectly by artificial intelligence "trained" by a certified operator (Papadakis and Mack, 1997).

Although other nomenclature such as *nondestructive evaluation* and *nondestructive inspection technology* are in use, NDT is the recognized generic nomenclature. Since NDT is such a results-oriented and ad hoc interdisciplinary field, it is appropriate to focus on NDT instruments while explaining methods. Better instruments in both standard and new subfields are coming out every year, so some equipment mentioned will seem obsolete even on the date of publication. A broad-brush approach will be used to keep the information current for as long as possible.

NDT customers (in particular, you the reader) are the users of NDT equipment. This book is concerned with justifying NDT in manufacturing for 100% of production. The NDT examples will be directed toward manufacturers. The investment in NDT equipment is not trivial by any means. One engineer at a manufacturer of aircraft claimed to have $4,000,000 worth of ultrasonic transducers (sensors, probes, and search units) in an array of drawers in a laboratory area (at 1971 economics, when a fully loaded midsize station wagon could be purchased for $4500). He wanted to buy some new ones of a particular external shape to fit into a groove in an aircraft part. The transducers were just the probes for multimillion-dollar systems.

NDT for objects is analogous to medical diagnostic ultrasound, x-rays, and MRI for the human body. As such, medical technology is much better known to the general public than is NDT. People may have their own bodies tested but not realize that the brake calipers in the cars they drive to the supermarket and the wing spars of the planes they take to far cities are tested also. The customer of the medical manufacturer may be the hospital, but the visibility of the doctor to the medical end user is much higher than is the visibility of the technician in the hangar of the major airline, for instance. Yet the airplane is stripped down to its bare bones at a D-Check every 4 years. On the other hand, I was impressed in 1967 that a mechanic in a car dealership used a dye penetrant (one NDT Big 5 Method) to prove that the cylinder head of my car was cracked. (This was maintenance, not manufacturing, but it is a human-interest example. It is not known whether the crack existed at manufacture and was exacerbated by road conditions.)

NDT can get close to the end user beyond the NDT customer. To focus ideas, my cocktail party response to the "Gee whiz, what is NDT?" question is that NDT does for airplane wings what your dentist does for your teeth with bite wings—finds the holes. Simplistic, but expressive. It should be comforting to the end user that the NDT expert will apply NDT for safety and can prove to his superiors that NDT should be used to improve the quality of the product he or she buys.

8.2 Various Classes of Methods: NDT and Others

8.2.1 Ultrasound

8.2.1.1 General View of Ultrasound in NDT

Ultrasound is sound above the range of human hearing. Ultrasound in NDT is an active radiation method, meaning that there is a source of ultrasound sending ultrasonic energy into the object being tested. It is mechanical radiation (Lindsay, 1960) analogous to infrared radiation (IR), light, ultraviolet, x-rays, and gamma rays, which are electromagnetic radiation. While electromagnetic radiation travels in free space and penetrates materials as is well known, mechanical radiation (ultrasound) travels in materials, namely gases, liquids, and solids. The ultrasonic radiation is then received, at least in part, by a receiver after traversing the object in a preassigned path. The resulting sequence of signals is displayed or processed for some kind of synthetic display or decision mechanism.

8.2.1.2 Production and Reception of Ultrasound

Consider the most generic type of ultrasonic radiating element. This is a piezoelectric plate with electrodes on both sides. Piezoelectric materials expand or contract (or else shear) depending on the direction of the applied voltage. If they experience a stress, they develop an electric charge, which is read by circuitry as a voltage. In other words, the piezoelectric elements can be used as transmitters and receivers for stress waves. The piezoelectric plate may be typically 0.5 inches (1.27 cm) in diameter and several thousandths of an inch thick (a fraction of a millimeter). The thickness defines half a wavelength of the ultrasound to be generated if the plate vibrates in a free-free bulk mode (not glued to anything). The wavelength is in the material of the piezoelectric plate, of course, and is related to the ultrasonic frequency, f, and the ultrasonic (mechanical wave) velocity v in the piezoelectric material by

$$\lambda = v/f \tag{8.1}$$

The piezoelectric plates are cut from piezoelectric crystals or are formed from ferroelectric ceramics that are poled (electrically polarized) in the proper directions. Poled ferroelectrics become piezoelectric, making them useful in linear acoustics. The useful cuts and directions are specified for two types of waves, longitudinal and shear (transverse). Longitudinal plates vibrate with particle motion in the thickness direction and generate longitudinal waves propagating normal to their major faces (see Figure 8.3). Shear plates, on the other hand, vibrate with particle motion in one direction in

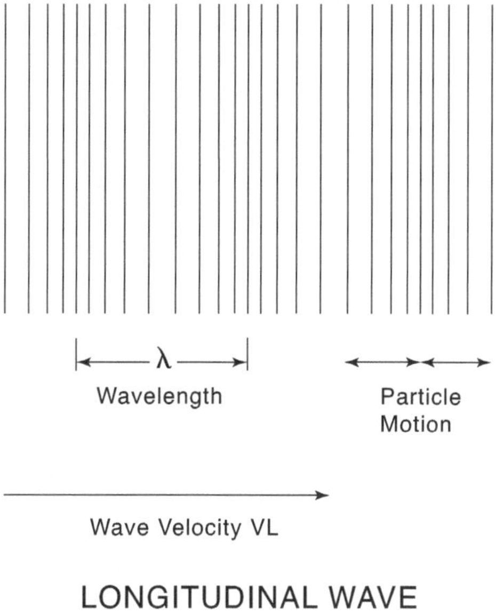

FIGURE 8.3
Longitudinal wave directions of propagation and particle motion. The strain is actually on the order of 1/1,000,000. (From Papadakis, E.P., ed. (1999). *Ultrasonic Instruments and Devices: Reference for Modern Instrumentation, Techniques, and Technology*, Academic Press, San Diego, CA, pp. 193–274. With permission.)

the plane of the major faces and generate shear waves also propagating normal to their major faces (see Figure 8.4). To produce ultrasonic beams from such plates, the lateral dimensions must be many wavelengths. For more details concerning piezoelectricity and piezoelectric plates, see Berlincourt et al. (1964), Cady (1946), Institute of Electronics and Electrical Engineers (IEEE, 1987), Jaffe and Berlincourt (1965), Jaffe et al. (1971), Mason (1950), Mattiatt (1971), and Meeker (1996).

In NDT, the term *transducer* refers to piezoelectric plates with backing and frontal elements to modify their vibration characteristics. These assemblies are potted inside cases to protect them and provide means for gripping them by hand or for mounting them in systems. The vast majority of ultrasonic NDT transducers are longitudinal (one design used extensively is shown in Papadakis et al., 1999).

Beams from transducers spread to some degree (Papadakis, 1991) as illustrated in Figure 8.5 (a very thorough summary of this phenomenon is given in Papadakis, 1975). Beam spreading affects both scientific and engineering uses of ultrasound. The spreading can be corrected for, sometimes rigorously and sometimes approximately. In NDT, the amplitude

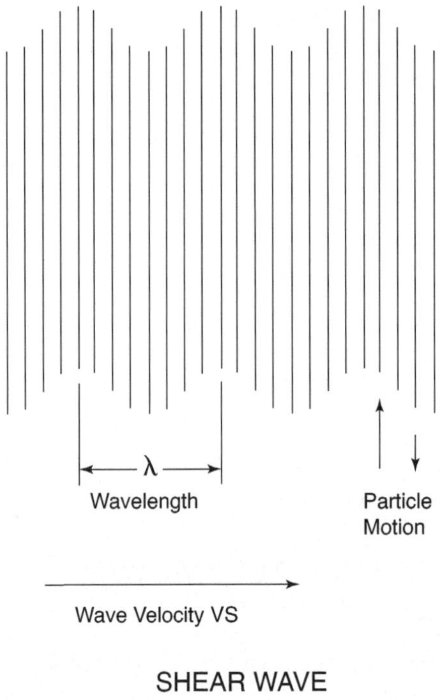

FIGURE 8.4
Shear wave directions of propagation and particle motion. The strain is actually on the order of 1/1,000,000. (From Papadakis, E. P., ed. (1999). *Ultrasonic Instruments and Devices: Reference for Modern Instrumentation, Techniques, and Technology,* Academic Press, San Diego, CA, pp. 193–274. With permission.)

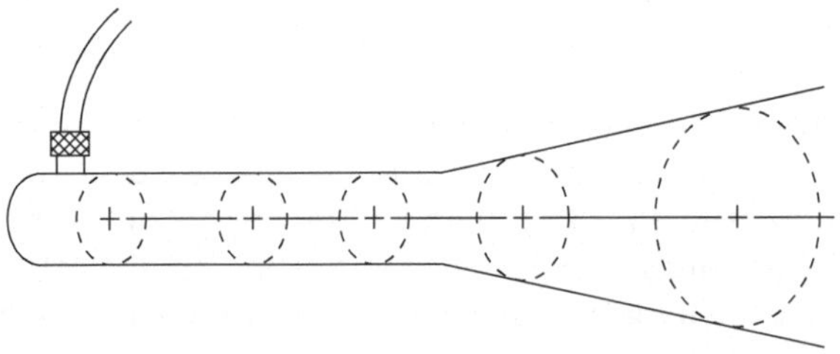

FIGURE 8.5
Schematic representation of spreading of an ultrasonic beam from a transducer. The ultrasonic wave is reflected by obstacles in its path. (From Papadakis, E. P. (1991). "Ultrasonic Testing." In *Nondestructive Testing Handbook*, 2nd ed., Vol. 7, Section 3, Part 5, eds. A. S. Birks, R. E. Green, Jr., and P. McIntire, American Society for Nondestructive Testing, Columbus, OH, pp. 52–63. With permission.)

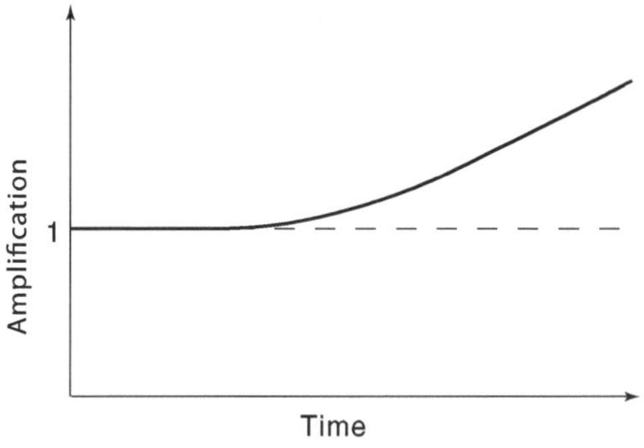

FIGURE 8.6
Schematic representation of the amplitude-distance-correction built into integrated flaw detection instruments as variable amplification applied to the returning echo signals. The ADC is an approximate beam-spreading correction. (From Papadakis, E. P. (1991). "Ultrasonic Testing." In *Nondestructive Testing Handbook*, 2nd ed., Vol. 7, Section 3, Part 5, eds. A. S. Birks, R. E. Green, Jr., and P. McIntire, American Society for Nondestructive Testing, Columbus, OH, pp. 52–63. With permission.)

of signals is sometimes corrected for distance approximately by a factor called ADC, the amplitude-distance-correction. The ADC depends on frequency, distance, piezoelectric plate diameter, and the velocity in the material supporting propagation. The ADC is electronically built into flaw detection instruments as amplification that varies with time (see Figure 8.6). Integrated instruments and display modes will be treated in the next section.

8.2.1.3 Integrated Instruments and Display Modes

8.2.1.3.1 The Generic Ultrasonic Instrument

Ultrasonic instruments could be set up in the laboratory using a multiplicity of components, each being a black box connected to others by cables. Indeed, most laboratories have such test sets that can be modified for development work. The typical bench-top test set (Papadakis, 1997a) could look like Figure 8.7. The synchronizing generator would typically be emitting 500 to 1000 pulses per second. The pulser could be emitting spike voltages or radio-frequency (RF) waveforms in the megahertz range. The pulse limiter keeps the pulser from overloading the amplifier while applying the full pulse voltage to the transducer and letting the small-amplitude echoes from inside the specimen, and from its back wall, go to the amplifier. The piezoelectric transducer in this picture is acting as both transmitter and receiver. The display would typically be a cathode ray oscilloscope. The computer is optional but is becoming ubiquitous.

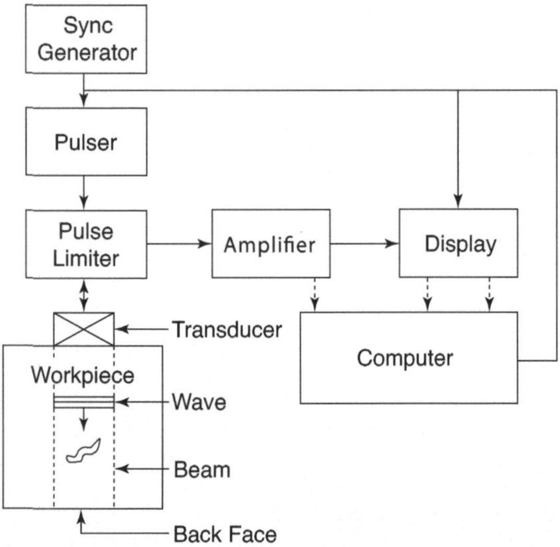

FIGURE 8.7
The typical bench-top test set consisting of synchronizing generator, pulser emitting spike voltages or RF waveforms, pulse limiter to keep the pulser from overloading the amplifier, piezoelectric transducer, CRO display, and computer (optional). (From Papdakis, E. P. (1997a). "Ultrasonic Instruments for Nondestructive Testing." In *Encyclopedia of Acoustics*, Vol. 2, ed. Malcolm J. Crocker, John Wiley & Sons, New York, pp. 683–693. With permission.)

8.2.1.3.2 The A-Scan Display

The display is termed an A-scan when the voltage is shown vertically and time is shown horizontally on the oscilloscope. See the stylized oscilloscope (Papadakis, 1997a) in Figure 8.8. With a broadband transducer and amplifier, and a negative spike for the pulser output, the picture would look like Figure 8.8(a). These signals, when rectified and detected, look like Figure 8.8(b). The rectifying and detecting circuit would be inserted between the amplifier and the display in Figure 8.7. It is not shown in the laboratory table-top test set because it is generally only used in integrated portable NDT ultrasonic instruments to make the display simpler and cheaper. For most flaw-detection applications, the rectified and detected signals suffice.

8.2.1.3.3 The Commercial Instrument

While size does not matter so much in the factory, customers want hand-portable flaw-detection instruments for the field. Of course portability is useful in the factory, also. One combined factory/field operation could be mentioned. Warships are built *in situ* at a seaport, and are hence their own factory in the field. Some ultrasonic flaw-detection instruments were specified to fit down the hatch in the conning tower of a submarine for use inside. In any case, robust integrated instruments combining all the necessary parts shown in Figure 8.7 plus other features are for sale by several manufacturers.

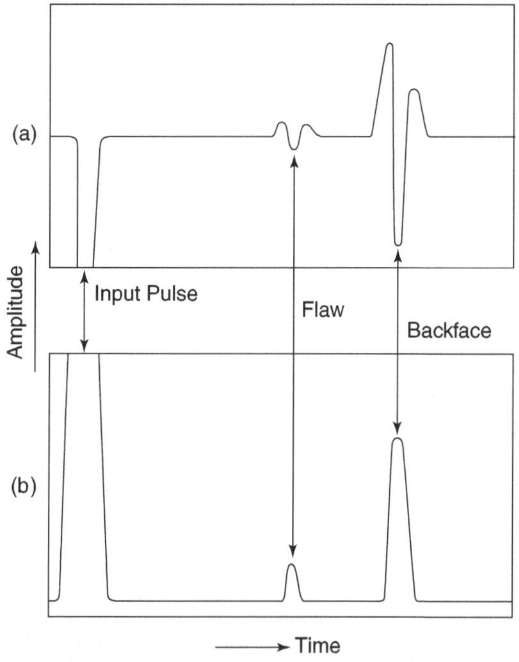

FIGURE 8.8
A-scan display has voltage vertically and time horizontally on the oscilloscope. (a) With a broadband transducer and amplifier, and a negative spike for the pulser output. (b) These signals when rectified and detected. (From Papadakis, E. P. (1997a). "Ultrasonic Instruments for Nondestructive Testing." In *Encyclopedia of Acoustics*, Vol. 2, ed. Malcolm J. Crocker, John Wiley & Sons, New York, pp. 683–693. With permission.)

All the individual black boxes shown in Figure 8.7 are interconnected in cases smaller than a cigar box or as big as a carry-on suitcase, depending on many parameters and specifications. The only external cable goes to the transducer.

Modern commercial instruments will not be enumerated because there are so many of them; they can be located through advertising in industrial magazines, particularly *Materials Evaluation*, the journal of the American Society for Nondestructive Testing. The buyer's guide in the June issue of the magazine every year lists all sorts of NDT instruments, manufacturers, and service organizations. Any NDT ultrasonic instrument currently in production can be found there. However, for interest, a few older instruments are mentioned here to show that the technology was in use constructively long before TQM and SPC tried to tear down the dependence on mass inspection.

In the nondestructive testing chapter of *Ultrasonic Instruments and Devices* (Papadakis, 1999), one will find in Figure 12 a photograph of a 1942-vintage ultrasonic flaw-detection instrument. Now at the University of Michigan, it belonged to one of the large aircraft manufacturers during the Second World War. Much good science and advanced NDT was done with this instrument at Michigan (Firestone, 1945a, 1945b; Firestone and Frederick 1946). The plan in Figure 8.2 was carried out with this instrument as an output and a tool. The instrument's oscilloscope, mounted ergonomically for a worker seated on a stool and manipulating a transducer manually, is of the vintage of radar display scopes used at Pearl Harbor, but misinterpreted with catastrophic consequences. The NDT of the airspace around the base was flawed. By contrast, the oscilloscope in the NDT instrument was used to good advantage to find flaws in aircraft materials that could have destroyed the aircraft without a shot's being fired. The dependence upon NDT to find latent defects is essentially total.

Another instrument mounted for factory use is shown in Figure 8.9. Its use will be described later as a special application of ultrasound to NDT.

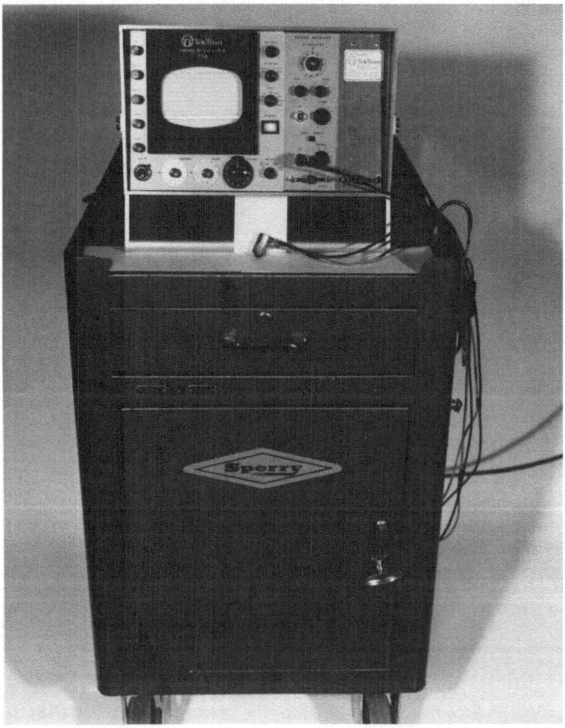

FIGURE 8.9
An ultrasonic flaw-detection instrument for factory use mounted on a dolly on casters. The dolly contains a water reservoir and a pump to force water into a perforated bladder on the front of the transducer. This structure facilitates coupling of the ultrasound from the transducer to the work piece. (From Papadakis, E. P. (1976b). "Ultrasonic Velocity and Attenuation: Measurement Methods with Scientific and Industrial Applications." In *Physical Acoustics: Principles and Methods*, Vol. XII, eds. W. P. Mason and R. N. Thurston, Academic Press/Harcourt, Inc., New York, pp. 277–374. With permission.)

This instrument, circa 1970, belies what was said before by having two attachments to the transducer, the electrical cable and a water tube. The water goes from a pump and reservoir mounted in the dolly on casters to a perforated bladder on the front of the transducer. This structure facilitates coupling of the ultrasound from the transducer to the work piece. The most usual coupling fluid for ultrasonic hand-held transducers is a hypoallergenic gel.

8.2.1.3.4 The C-Scan Display

Let us assume you want to ensure that there are no flaws larger than a certain minimum size in a piece of metal to be machined into a critical part such as the wing spar of an airplane, and suppose further that you wish to minimize machining expenses on faulty material. In other words, you want to inspect raw material such as a thick rolled plate of aluminum for flaws. An automatic system can be assembled with an ultrasonic NDT flaw-detecting instrument and some extra circuitry and computers to scan the entire interior of the plate. The system is set up with the plate and the transducer in a water bath. The transducer is carried on a gantry to scan over the entire plate. The timing of the electronic gates letting through the received echoes is set to eliminate large echoes from the surfaces of the plate and detect only echoes from flaws in the interior. A generic picture of a C-scan is given in Figure 8.10. Suppose the gantry sweeps the transducer along the x-axis all the way across the part and then returns. Upon the return, the gantry advances a small increment along the y-axis and repeats the sweep across x and back. While the gantry is moving across x, the transducer is pulsed many times to send waves through the water, through the part, and back again to the transducer as receiver. The speed of traverse along x is regulated such that many pulses of ultrasound enter the part to detect all the flaws in it. After the traverse along y is completed, the ultrasound has prepared a picture of the entire interior of the work piece.

8.2.1.4 Specialized Instruments and Applications

8.2.1.4.1 Large C-Scan for Flaw Detection in Airplane Wings — Probability of Detection

8.2.1.4.1.1 The System — To show the magnitude of the facilities built for specialized NDT operations (Papadakis, 1997a), the C-scan in an aircraft factory is shown in Figure 8.11. The tank in which the plate to be tested is immersed is as big as two lanes of an Olympic swimming pool. In actuality, this is a rather small system, as wings of large commercial jetliners require more space than this. In the pictured system, the y-axis from Figure 8.10 is along the length of the tank, and the x-axis is along the width of the tank. The moving gantry stretches the width of the tank and is at the center of the picture, slightly to the right of the rack of instrumentation with oscilloscopes. The resulting output is a picture of the interior of the test piece showing

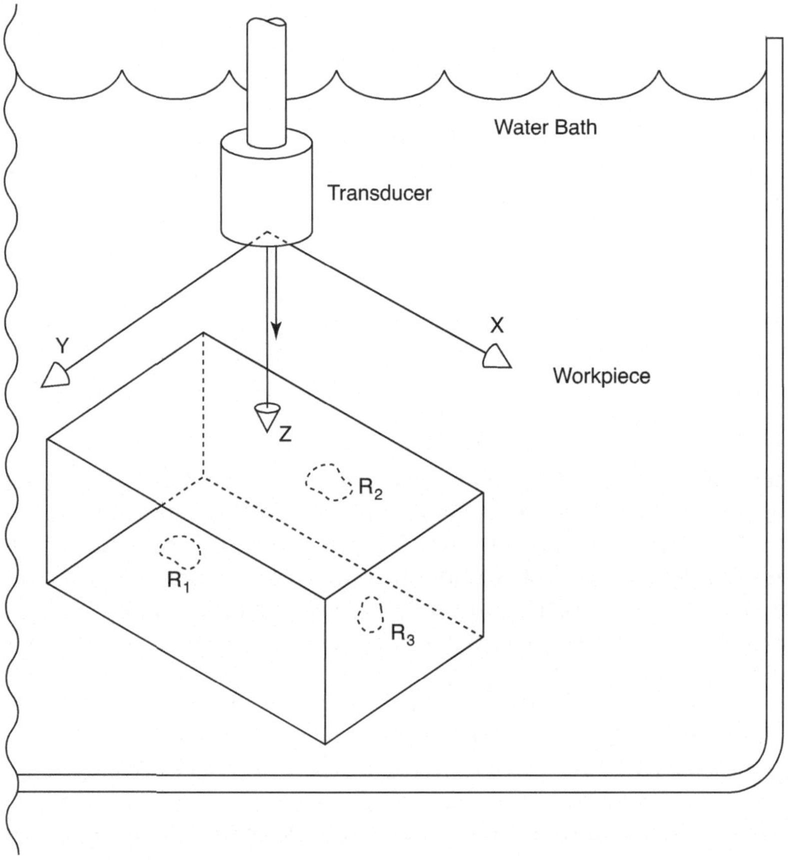

FIGURE 8.10
A generic picture of a C-scan. The operation is described in the text. (From Papadakis, E. P. (1997a). "Ultrasonic Instruments for Nondestructive Testing." In *Encyclopedia of Acoustics*, Vol. 2, ed. Malcolm J. Crocker, John Wiley & Sons, New York, pp. 683–693. With permission.)

flaws as ultrasonic echoes, which are made visible electronically. This is like looking for cysts on a kidney or genitalia on a fetus.

8.2.1.4.1.2 Probability of Detection — Giant C-scans like these, and smaller ones that may use (r, θ) coordinates as well as (x, y) coordinates, are used principally for flaw detection. In this regime the concept of probability of detection (POD) is of critical importance. With the human eye looking at an oscilloscope, a signal amplitude analyzer testing the voltage reading from the echo, or some sort of artificial intelligence examining the echo from a suspected flaw, there is a range of echo sizes that are ambiguous. The prime fact to be understood is that some flaws are so small that they will not be detrimental. That is, they will not grow enough under cyclical stresses to cause failures before the next inspection. In the aircraft case, the stresses are on takeoff,

FIGURE 8.11
The C-scan in an aircraft factory. The tank is as large as two lanes of an Olympic swimming pool; other systems are still larger. In the pictured system, the y-axis from Figure 8.10 is along the length of the tank and the x-axis is along the width of the tank. The moving gantry carrying the transducers stretches the width of the tank and is at the center of the picture. (From Papadakis, E. P. (1997a). "Ultrasonic Instruments for Nondestructive Testing." In *Encyclopedia of Acoustics*, Vol. 2, ed. Malcolm J. Crocker, John Wiley & Sons, New York, pp. 683–693. With permission.)

landing, and in unexpected maneuvers. The next inspection would be scheduled soon enough so that the growing flaws would not have grown to criticality prior to the inspection. The inspection cycle is determined by the industry and the Federal Aeronautics Administration (FAA). This inspection would be maintenance, not the inspection during manufacturing. The next inspection might be done with a handheld probe on a portable test set rather than by a big C-scan. In each case, the probability of detection of the inspection method would need to be of the necessary sensitivity to see the size flaws in question.

A schematic graph of the POD is shown in Figure 8.12. Here the S-curve labeled *real technique* is the POD curve. The sensitivity of the technique is adjusted such that the critical flaw size, or in some cases a flaw size smaller than critical, which is to be allowed, falls within the S-curve of the POD. Such a case is illustrated here. Then there is a fraction accepted (FA) of flaws, which are larger than desired, and a fraction rejected (FR) of flaws, which are smaller than permissible. For a good test, both of these fractions are small. The test may not be symmetrical; the sensitivity may be set for a very small FA while permitting a moderate FR. FA is a question of safety; FR is a question of cost.

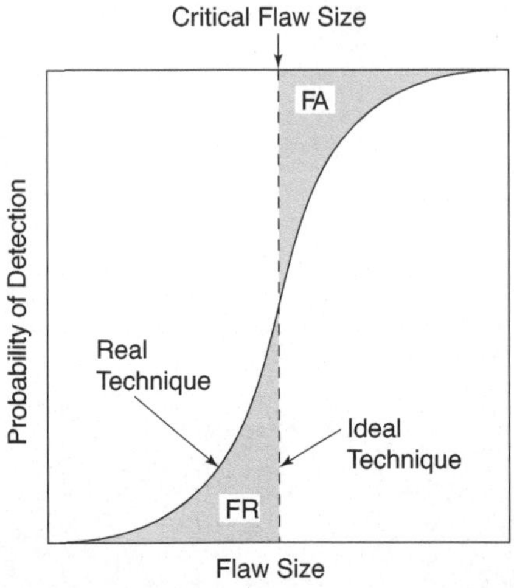

FIGURE 8.12
A schematic graph of the probability of detection (POD). The S-curve labeled Real Technique is the POD curve. The sensitivity of the technique is adjusted such that a flaw size that is to be allowed falls within the S-curve of the POD. A fraction, false accepts (FA), of flaws larger than desired are accepted and a fraction, false rejects (FR) of flaws that are smaller than permissible are rejected. For a good test, both of these fractions are small. The test may not be symmetrical; the sensitivity may be set for a very small FA while permitting a moderate FR. (From Papadakis, E. P. (1992). "Inspection Decisions Based on Costs Averted," *Materials Evaluation*, 50(6) 774–776. With permission.)

POD curves are arrived at empirically by measuring whether or not actual flaws of different sizes are detected by operators using the detection means. Doing a thought experiment, one might find that 25% of a set of operators found a surface-breaking crack 1.00 mm long 90% of the time, 50% of the operators found a surface-breaking crack 1.25 mm long 90% of the time, 75% found one 1.50 mm long similarly, 95% found one 2.00 mm long, and so on. POD curves are not limited to ultrasonic echoes but are applicable to dye penetrants, magnetic particles, eddy currents, and x-rays as well. It is probable that a POD can be concocted for any inspection method.

8.2.1.4.2 Immersion Tank and System for Automotive Nodular Iron Parts Strength

8.2.1.4.2.1 Strength and Graphite Shape — Nodular iron parts must be tested for nodularity (percent of the graphite in spherical particles) as well as for flaws. More details will be given in Section 8.3 on correlations and functions as well as in Chapter 9 on specific financial calculations on examples. Photomicrographs illustrating nodularity are given in Figures 9.1(a) and (b). For maximum strength, one wants the minimum of any shape of graphite except spheres. It turns out that the maximum amount of free graphite in spheres leads to the maximum ultrasonic velocity. The reason is that the strong iron is maximally connected around the spheres, whereas it is cut up more by the other shapes of weak graphite. Hence, one wants to measure ultrasonic velocity and set a reject limit at some high attainable ultrasonic velocity to ensure adequate strength of the iron. We are considering here the measurement of intrinsic variables for material properties.

8.2.1.4.2.2 *Generic Velocity Tank Diagram* — A generic drawing of the ultrasonic tank for this sort of measurement is given in Figure 8.13. The input and output transducers IN and OUT are situated in the water *W* at a distance *L* apart. The transit time t_0 is measured. Then the metal *M* is brought into place, and the two other transit times t_1 and t_2 for the paths shown are measured. (The path for t_2 is drawn displaced vertically from t_1 for clarity only.) The three times are sufficient to calculate the path *L*, the length *d*, and the velocity *v* in the metal.

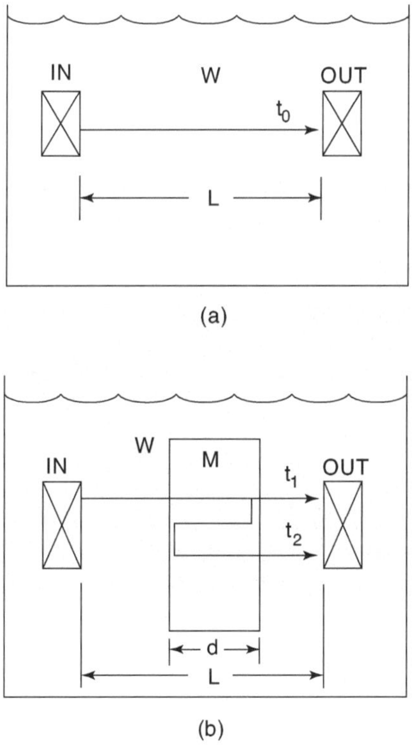

FIGURE 8.13
Generic drawing of the ultrasonic tank for measurement of velocity in nodular iron. The input and output transducers IN and OUT are situated in the water *W* at a distance *L* apart. The transit time t_0 is measured. Then the metal *M* is brought into place, and the two other transit times t_1 and t_2 for the paths shown are measured. (The path for t_2 is drawn displaced vertically from t_1 for clarity only.) The three times are sufficient to calculate the path *L*, the length *d*, and the velocity *v* in the metal. (From Papadakis, E. P. (1976b). "Ultrasonic Velocity and Attenuation: Measurement Methods with Scientific and Industrial Applications." In *Physical Acoustics: Principles and Methods*, Vol. XII, eds. W. P. Mason and R. N. Thurston, Academic Press, New York, pp. 277–374. With permission.)

8.2.1.4.2.3 Strength vs. Velocity — A graph for the iron to be used in one part is shown in Figure 8.14. Strength is plotted vs. ultrasonic velocity. Both tensile strength and yield strength are shown. The band of values of each variable is the spread found empirically between the upper 95% confidence limit and the lower 95% confidence limit on variables, which have some statistical variability. The data consisted of measurements on some 150 tensile bars of different nodularity chosen optically. The percent nodularity

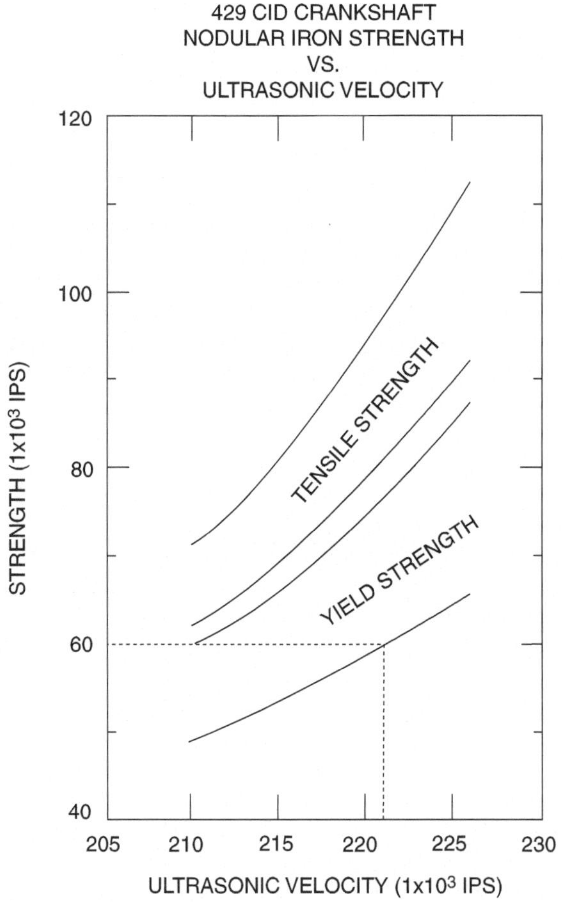

FIGURE 8.14
Tensile strength and yield strength for nodular iron are plotted against ultrasonic velocity. Ninety-five percent confidence limits are shown from data on some 150 tensile bars. The reject limit is drawn at 60,000 psi. Parts are designed with that minimum yield strength in mind for the iron. Where 60,000 psi intersects the lower 95% confidence limit on yield strength, the ultrasonic velocity v is 221,000 inches per second, so v must be higher than that for acceptance. (From Papadakis, E. P. (1976b). "Ultrasonic Velocity and Attenuation: Measurement Methods with Scientific and Industrial Applications." In *Physical Acoustics: Principles and Methods*, Vol. XII, eds. W.P. Mason and R.N. Thurston, Academic Press, New York, pp. 277–374. With permission.)

is the percentage of the free graphite estimated to be in the spherical form. The graphs in Figure 8.14 are distilled from empirical measurements of ultrasonic velocity in the tensile bars and from the results of pulling the tensile bars.

The research leading up to the discovery that ultrasonic velocity as a predictor of strength in nodular cast iron was not straightforward (Torre, 2005). In the early days of nodular iron, around 1970 when it was first being suggested as a substitute for steel in stressed parts, Rocco Torre, a salesman for Sperry Instruments, was working with Milt Diamond and Bob Lutch, engineers at General Motors. Sperry sold ultrasonic pulse-echo equipment. Rocco, as a sales engineer, was attempting to correlate nodular iron quality with ultrasonic attenuation, which is the rate of dying out of ultrasonic pulses as they travel through a material. Attenuation had recently been shown (Papadakis, 1964) to be a sensitive indicator of heat treatment results in steel (see Figure 8.15). (The iron portion, about 86%, of cast nodular iron is essentially steel.) The attenuation method for nodular iron strength showed technical feasibility but was labor intensive including a skill component. It was not clear that the method would ultimately pass plant feasibility. In the process of studying attenuation in nodular iron, the stability of the time base on the Sperry Reflectoscope became suspect. Torre discovered that the time between echoes actually varied from sample to sample of the same size; the Reflectoscope was indeed stable. Further work with Jerry Posakony of the Sperry home office using flat and parallel specimens with better size specifications proved that the velocity in the iron varied monotonically with the nodularity of the iron and hence with the strength. An automatic velocity measurement system based on the tank sketched in Figure 8.13 was constructed and installed commercially at the GM plant in Defiance, Ohio. The process of technology transfer and commercialization emphasized by the author (Papadakis, 1999) was speedy. Various technical improvements were made in rapid succession. Thus, by serendipity, Torre and the others discovered that ultrasonic velocity was the best way to ensure nodular iron strength. Torre worked with GM, Ford, and Chrysler for many years, and participated in the Detroit chapter of the ASNT.

8.2.1.4.2.4 Reject Limit and Equivalent POD — The reject limit mentioned above is drawn at 60,000 psi in Figure 8.14. Parts are designed with that minimum yield strength in mind for the iron in this case. Where 60,000 psi intersects the lower 95% confidence limit on yield strength, the ultrasonic velocity is 221,000 inches per second (0.221 in/μsec). Thus, one wants iron with ultrasonic longitudinal velocity greater than or equal to 0.221 inches per microsecond. The common English engineering units for stress are psi and in/μsec for ultrasonic velocity. Nodular iron near its essentially maximum velocity of 0.225 in/μsec is easily obtainable by factory metallurgical treatment. Analogous to the POD, there will be a few false accepts falling below 60,000 psi, but to the right of 0.221 in/μsec. In practice to be conservative, the reject limit is often set as high as 0.224 in/μsec. (Melting and

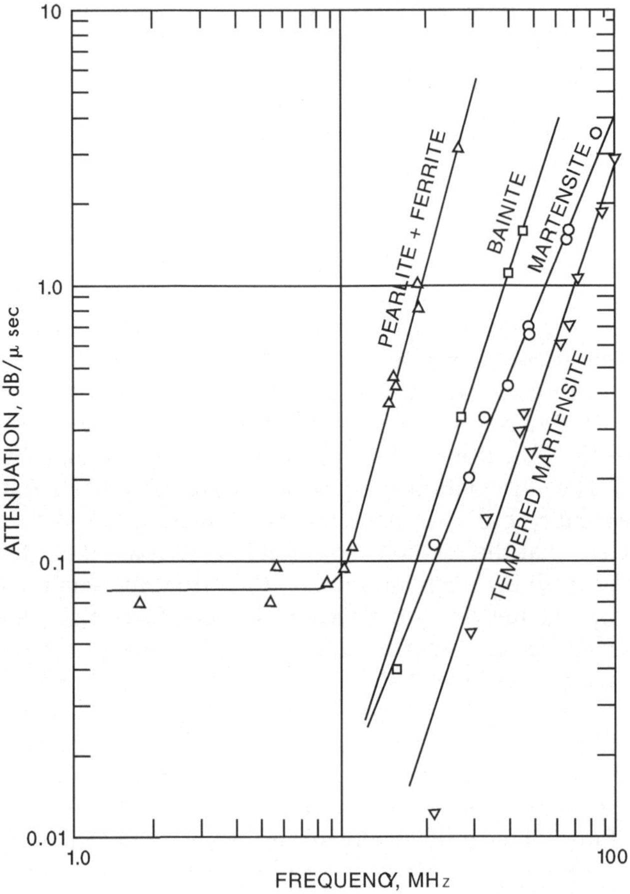

FIGURE 8.15
Ultrasonic attenuation in three transformation products in steel quenched at different cooling rates from the same austenitizing temperature. Also shown are the results of tempering the hard brittle martensite. The high slopes on log-log paper indicate a large effect of grain scattering by the prior austenite grain volumes subdivided by the microstructures. (From Papadakis, E.P. (1964). "Ultrasonic Attenuation and Velocity in Three Transformation Products in Steel," *Journal of Applied Physics*, 35, 1474–1482. With permission.)

recasting a few extra false rejects is inexpensive compared with the cost of an accident.) Ultrasonic velocity tests are performed at iron foundries and product-oriented casting plants.

8.2.1.4.2.5 Large Practical Tank — A tank holding fixtures and transducers for testing a right and a left front wheel spindle support is shown in Figure 8.16. The spindle is the stubby axle on the front of a car, where there is one on each side. The spindle support is attached to the McPherson strut and has the steering push rods and the brake calipers attached to it.

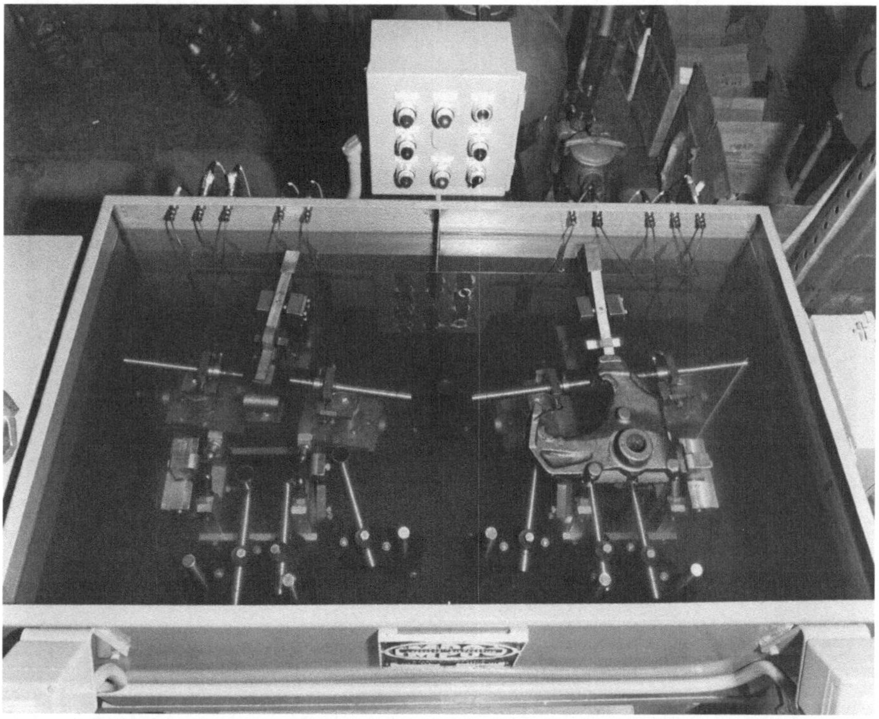

FIGURE 8.16
A tank holding fixtures and transducers for testing a right and a left nodular iron front wheel spindle support. Certain areas are inspected for flaws while velocity is measured in another area for strength assurance. The left fixture is empty while the right fixture holds a spindle support ready for testing. The pairs of horizontal and horizontally opposed coaxial wands (stainless tubes) hold transducers for the velocity measurements. On the right, the part of the spindle support to be measured is between the transducers. On the left, a calibration block is between that pair of transducers. On each side, three other transducers for flaw detection face upward in the water at the ends of other wands. Their faces are the jet black discs in the left picture. (From Papadakis, E. P. (1976b). "Ultrasonic Velocity and Attenuation: Measurement Methods with Scientific and Industrial Applications." In *Physical Acoustics: Principles and Methods*, Vol. XII, eds. W.P. Mason and R.N. Thurston, Academic Press, New York, pp. 277–374. With permission.)

Its integrity and strength must be ensured, as it is a critical safety part. In operation, this tank would be manually loaded and unloaded with pairs of spindle supports.

8.2.1.4.2.6 The Electronics for the Tank — In Figure 8.17 the rack of electronics, circa 1975, stands next to the tank in front. The electronics were interconnected to measure the velocity in each piece and the flaws in three areas of each piece, and to read out and record the results. When the two parts were loaded and seated properly, the two-handed Occupational Health and Safety Administration (OSHA) switches on the front of the tank were pushed simultaneously by the operator, and the automatic electronics functioning

FIGURE 8.17
Automatic rack of electronics to measure the velocity in each piece and the flaws in three areas of each piece stands next to the tank of Figure 8.16. The two parts were loaded and the two-handed OSHA switches on the front of the tank were pushed simultaneously by the operator to start the electronics. (From Papadakis, E. P. (1976b). "Ultrasonic Velocity and Attenuation: Measurement Methods with Scientific and Industrial Applications." In *Physical Acoustics: Principles and Methods*, Vol. XII, eds. W.P. Mason and R.N. Thurston, Academic Press, New York, pp. 277–374. With permission.)

commenced. The system is considered to be semiautomatic. Other systems have automatic (sometimes robotic) loading and unloading. One ingenious system with automated materials handling was fitted with a drill to bore a needed bolt hole in only those parts that passed the test (Klenk, 1977). As no other automation would bore the hole, a flawed or nonconforming part could not be assembled on a car even if it were shipped to the assembly plant inadvertently or by a manufacturing manager wishing to meet his quota.

8.2.1.4.3 Water Column Transducer on Flaw Detector for Spot Weld Assurance

8.2.1.4.3.1 Flaw Detection Instrument — A portable ultrasonic flaw detection instrument was shown in Figure 8.9. As described there, it is mounted on a dolly with a water pump in the lower section. That arrangement is necessary for the spot weld quality assurance. This technology was reported earlier (Papadakis, 1976b). The pump and a reservoir supply water to a plenum behind a perforated membrane mounted on the front of the transducer.

The water column in the plenum permits the transducer to send ultrasound pulses into the spot weld as the membrane touches the accessible side of the spot weld. Ultrasound echoes back and forth in the spot weld, a portion of which comes back into the water column at each echo. The series of echoes in the spot weld are analyzed to determine the size and quality of the spot weld nugget. The analysis will be described after the description of the spot welds and the transducers.

8.2.1.4.3.2 *Spot Welds* — Spot welds are made by passing a high electric current through two sheets of metal clamped together by electrodes. Generally the electrodes are copper with high electrical and thermal conductivity. Starting at the interior boundary between the two sheets of metal to be welded, the current begins to melt the metal. The electrical current is allowed to flow long enough to melt a region about as wide as the electrodes and almost as thick as the two sheets of metal. Then the current is turned off and the electrodes remain clamped long enough for the molten region to recrystallize into a nugget of solid metal of coarser grain size. A well-formed nugget is shown in Figure 8.18. One will notice that the grains are larger than in the parent metal and that they are columnar, growing in from the positions of the cool electrodes. The layman can see spot welds on car door jambs, on train passenger cars, and on stainless steel teapots in Chinese restaurants.

Specifications for spot welds in the automotive industry are written on the basis of a tear-down test. Actual cars and car parts are ripped apart with jackhammers and crow bars. The parent metal, not the nugget, must rip (Ford Motor Co., 1972). The nuggets must be of a certain size, say 7 out of 10 in a row, with only two smaller than the specified size, and only one

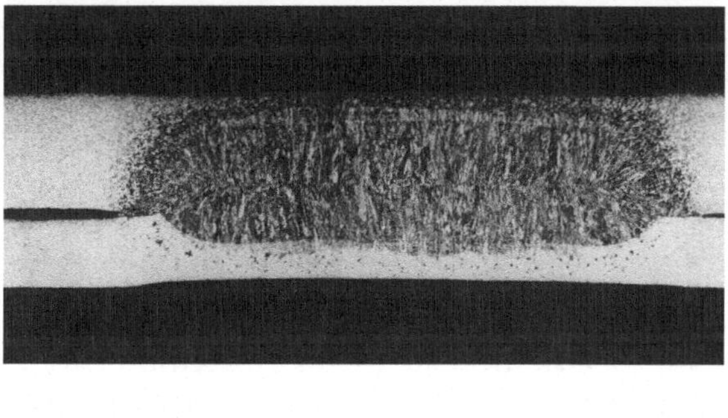

FIGURE 8.18
Section cut through a spot weld connecting two sheets of steel. The diameter and thickness of the nugget are important for strength. These dimensions are measured by ultrasonic echoes. (From Papadakis, E. P. (1976b). "Ultrasonic Velocity and Attenuation: Measurement Methods with Scientific and Industrial Applications." In *Physical Acoustics: Principles and Methods*, Vol. XII, eds. W.P. Mason and R.N. Thurston, Academic Press, New York, pp. 277–374. With permission.)

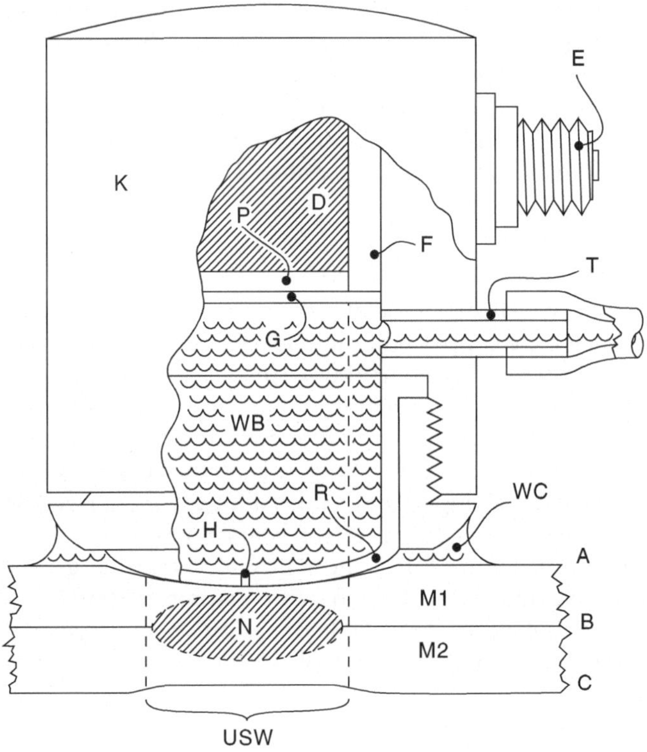

FIGURE 8.19
Spot weld transducer cutaway view. (From Papadakis, E.P. (1976b). "Ultrasonic Velocity and Attenuation: Measurement Methods with Scientific and Industrial Applications." In *Physical Acoustics: Principles and Methods*, Vol. XII, eds. W.P. Mason and R.N. Thurston, Academic Press, New York, pp. 277–374. With permission.)

missing. Any ultrasonic NDT test must be able to predict this behavior to be considered valid.

8.2.1.4.3.3 Spot Weld Transducer — A cutaway view of a spot weld transducer is shown in Figure 8.19. The case, K, is small enough to be held by the thumb and two fingers. The electrical cable is attached at terminal E, which applies the electric field to the piezoelectric plate, P, between the bottom of the damping backing, D, and the top of the protective wearplate, G. F is a cylindrical spacer. The water is introduced through the tube, T, into the water bath, WB, inside the rubber membrane, R, perforated by the hole, H. Water flows out into the meniscus of the water column at WC, making the water continuous through the rubber, which is a good match to the water. The width of the ultrasonic wave from P is USW. This is centered by the operator to go through the nugget, N, between the two sheets of metal M1 and M2. An ultrasonic pulse is sent from P to surface A. There are reflections from A and multiple reflections between A and C with extra reflections from interface B if the nugget is undersized.

8.2.1.4.3.4 Analysis of Echoes — The first principle is that there are echoes from surfaces A and C, but none from B if the nugget is wide enough. The second principle is that the echoes in the nugget die out fast if the nugget is thick enough. This is because of the high attenuation in the recast metal in the nugget. A corollary of the first principle is that there is evidence of echoing from surface B if the nugget is undersized. In particular, if there is no nugget but only surface damage from a stick weld when the two layers are torn apart, then the echo pattern will show a great deal of echoing from surface B. The third principle is that there will be echoes between A and B alone of there is no weld at all. All these situations are illustrated in Figure 8.20, which shows echoes from samples that were subsequently torn down. The operator, an hourly employee on the production line, is trained to do this ultrasonic test, recognize the echo patterns, and judge the quality of the spot welds. This sort of a test can permit the salvaging of the three parts (up to whole car bodies) per hour when they are torn down to check whether the welds had been good.

8.2.1.4.3.5 Automation — Automation of this test has not been accomplished yet. Methods considered include phased array transducers in a water tank to aim the ultrasonic beam at the nugget along the normal to the sheet metal face. Artificial intelligence would be needed to perform the analysis of the echo pattern on the fly.

8.2.1.4.4 Water Bubbler Ultrasonic Assembly To Test for Chevrons in Forward-Extruded Axles

8.2.1.4.4.1 The Forward Extrusion Process — New processes bring new problems. In early times, a shaft would be turned on a lathe starting with a steel rod as raw material, the rod being larger in diameter than the finished shaft. Piles of chips would be made by the long cutting process. The shaft in final shape would then be heat treated.

The new process is forward extrusion. A die is made and hardened for each reduction in size of the shaft. The die has a chamber for the billet of raw material and a hole the size of the intended reduced diameter. The rim of the hole is rounded to facilitate the sliding through of the compressed steel. The raw material is a billet of steel, annealed soft, about the shape of a can of soup and large enough to contain the volume of the final shaft. A piston forces the billet into the hole, decreasing the diameter and extending

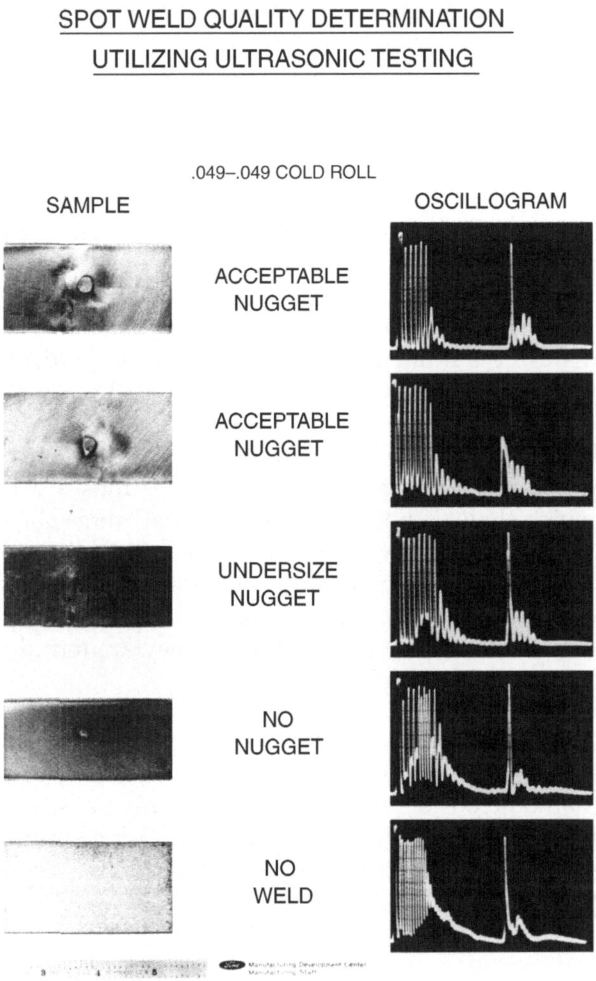

FIGURE 8.20
Tear-down showing nuggets and their echo patterns. (From Papadakis, E.P. (1976b). "Ultrasonic Velocity and Attenuation: Measurement Methods with Scientific and Industrial Applications." In *Physical Acoustics: Principles and Methods*, Vol. XII, eds. W.P. Mason and R.N. Thurston, Academic Press, New York, pp. 277–374. With permission.)

the length of the material forced into the hole. Several reductions in diameter may be made on one shaft by several dies. The resulting shaft is indistinguishable from the shaft made on the lathe except for the lack of tool marks. In this case, too, heat treating is then done. A few seconds of extrusion time are substituted for many minutes of lathe work. Stress risers from tool marks are eliminated. All the chips made by lathe work are saved.

As always with substitutions, there is a trade-off. The new process, forward extrusion, introduces the possibility of a new kind of flaw, the chevron.

This will be described in the next section. Then the process and equipment for detecting it will be explained.

8.2.1.4.4.2 The Chevron: A New Flaw — In the forward extrusion process, the metal is forced past the rounded shoulder into the smaller-diameter hole. In the hole it must elongate to maintain its density. Just inside the hole, the mechanics of the process of elongation are such that the metal is subjected to an internal tension stress along the centerline.

The original piece of metal is cut from a billet, and billets are made from ingots that are cast in foundries in molds. Ingots are notorious for having shrinkage along their centerlines at the top of the mold. Indeed, the top end of all ingots is sheared off in the foundry and put back into the melting furnace because of this deleterious shrinkage. The ingot may still have microscopic shrinkage that is reduced in diameter by the rolling process, which reduces the diameter of the ingot to that of the desired billet.

Shrinkage flaws along the centerline of the billet, when subjected to the tension stress inside the die, may produce cracks. If these cracks occur, they appear in the shape of a conical internal rip with its apex pointing in the direction of the extrusion like a spear point. If the shaft is cut in half along its centerline, the bisected rips in the metals look like chevrons, that is, sergeants' stripes. A section of an extruded shaft, cut thus, is shown in Figure 8.21.

From the point of view of safety, chevrons inside certain shafts, such as axles, are safety hazards. Thus, 100% inspection should be performed. From the point of view of quality control, these flaws are not just nonconforming material, but are actually flaws. From the process point of view, the process is never under control because there is no way to process or inspect the raw material to ensure no chevron formation in the tension environment of the forward extrusion. One hundred percent inspection should be performed here, as on all processes that are never under control. In actuality, a test is available and can be installed, as will be shown below.

8.2.1.4.4.3 Ultrasonic Test for Chevrons — Either x-rays or ultrasonic pulse-echo inspection could be used to detect chevrons in shafts. Ultrasound was chosen and summarized in an overview (Papadakis, 1980). The ultra-

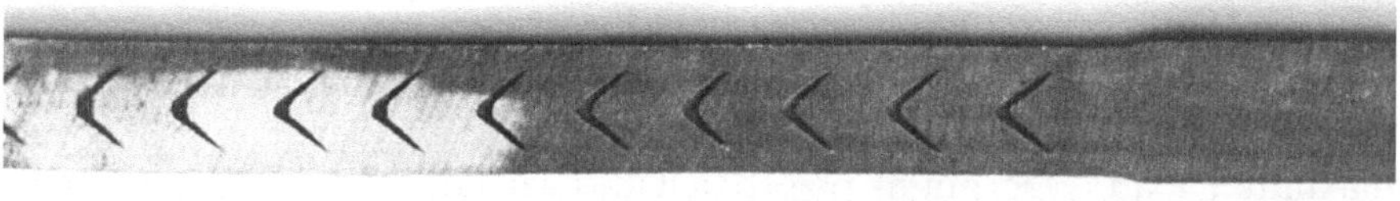

FIGURE 8.21
Photograph of chevrons in a forward-extruded shaft. The shaft has been sectioned lengthwise. Internal tension stresses in the extrusion process cause internal rips in the metal. (From Papadakis, E. P. (1981a). "Challenges and Opportunities for Nondestructive Inspection Technology in the High-Volume Durable Goods Industry," *Materials Evaluation*, 39(2), 122–130. With permission).

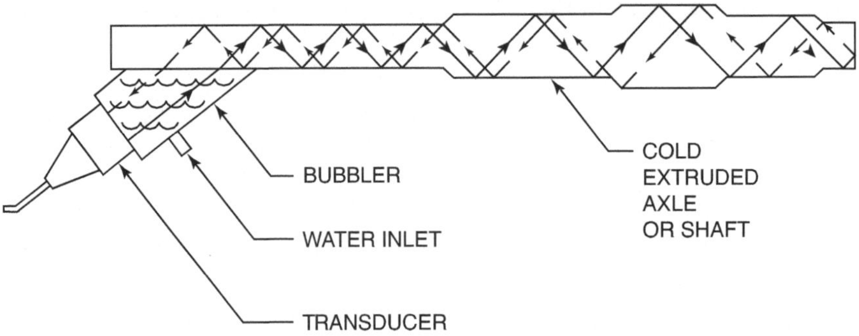

FIGURE 8.22
Diagram of ultrasonic water bubbler introducing ultrasonic waves into a shaft. If chevrons were present, extra echoes would appear at locations before the end of the shaft. As the forward-extrusion process may produce chevrons at random, 100% inspection by NDT is necessary. Ultrasound is faster and better in many ways than x-rays. (From Papadakis, E. P. (1981a). "Challenges and Opportunities for Nondestructive Inspection Technology in the High-Volume Durable Goods Industry," *Materials Evaluation*, 39(2), 122–130. With permission).

sonic pulse generated by a transducer is introduced into the shaft by a water bubbler as the coupling means. A diagram is given in Figure 8.22. Once inside the shaft, the ultrasonic pulse echoes down the shaft, reflects, and echoes back to the water bubbler and the transducer. If there are chevrons in the way, the ultrasound would echo from them and return to the transducer at an earlier time. Doing so, it would be detected and an alarm mechanism would be activated. The shaft would be removed from production and destroyed, ensuring no danger to a customer.

This vital test is one of the earliest automated ultrasonic inspections in the automobile industry.

8.2.2 Acoustic Emission (AE)

8.2.2.1 General View of AE in NDT

AE is a stress wave emitted at the tip of a crack as the crack propagates under stress. One deliberately applies a macroscopic stress and listens for the microscopic stress waves (mechanical radiation) that may be generated. The sound is generally in the frequency range of a few hundred kilohertz, although it can be audible as in the cracking of ice cubes when a cold drink is poured over them. The macroscopic stress would be larger than the stress the part would be expected to see in service, but smaller than the design maximum load. One might object that the causing of crack propagation makes the method destructive rather than nondestructive, but it is generally nondestructive in the same sense that proof-testing is nondestructive. The stressing to generate AEs is like adding one more fatigue cycle to a part that should be able to sustain several thousand fatigue cycles before failure. Astute analysis of the AE activity is necessary to determine whether a new part is fit for service or whether a used part should be repaired before being returned to service (in the maintenance mode).

8.2.2.2 Production and Reception of Acoustic Emission

As mentioned above, a macroscopic stress is applied to a part. The macroscopic stress may be pressure inside a pressure vessel, a compression or a tension in a tensile machine, or a torque. The sound is received by a piezoelectric transducer, which is usually a resonant type built in the frequency range expected. The resonance modifies the signal by building up many cycles of response, whereas the actual AE stress wave may be just a spike. Several papers on methods are collected in ASTM STP 505 (ASTM, 1972). While propagating from the source (crack) to the transducer, the signal may be spread out by mode conversion to look nothing like its original form (see Papadakis and Fowler, 1972). What is important is that the output from the transducer is proportional in some sense to the input from the crack propagation. Each motion of the crack tip results in a separate burst of acoustic emission. The burst is partially defined by the resonance of the transducer. However, one attempts to count the number of bursts occurring during some parametric interval such as the period of time while macroscopic stress is increasing. This counting is performed by means of an instrument consisting of an amplifier, some signal conditioning circuitry, and an electronic counter.

8.2.2.3 Integrated Instruments and Display Modes

Instruments are so specialized that it is not productive to show any particular instrument. In general one can say that the instruments are characterized by the number of channels of data handled. Use of several channels simultaneously permits the user to triangulate on the source of emissions in a large complex shape. This is termed *source location*. The position of the source is found in terms of coordinates established in the applications engineering phase of the AE project. Displays can be electronic, paper charts, and so on.

Single-channel instruments may be set up to accept or reject parts on the basis of the amount of AEs heard during a stressing routine. The level of AE to be considered dangerous to the performance of a part in later service must be determined by tests to failure. The display in this case is a count for each part.

8.2.2.4 Specialized Instruments and Applications

8.2.2.4.1 Instruments

The reader is referred to sales literature from manufacturers for information on specialized instruments. Sales literature can be traced through the buyer's guide in the June issue of *Materials Evaluation*, the journal of the ASNT.

8.2.2.4.2 An Experiment

One experiment holding a great deal of potential for the testing of brittle materials will be described here (Papadakis, 1981b). The statisticians in the audience will love this because it requires lognormal probability distributions to explain the results and find reject limits.

In the early days of research into the substrates for automotive catalytic converters, it was not clear whether the porous, thin-walled ceramics (needed for the afterburner system to oxidize unburned exhaust gases) would survive 50,000 miles of use. Tentative specifications caused worry about cracks. It seemed that cracks in a brittle material of a simple exterior shape, a cylinder, provided a good test for AE.

Specimens made under factory conditions were available in various degrees of completion before mounting in the housings to fit into the exhaust systems of cars. Large batches of each type were obtained for testing. It was desired to test them in compression and also in torque to determine which macroscopic stress might be better for a test procedure. A computer-controlled tensile machine was available that operated in compression as well as in tension. A fixture was constructed to hold the ceramic cylinders in the jaws for compression. A manual torque wrench with electronic output was fitted to this fixture to permit the torque measurements while the compression was held at maximum.

The fixture is shown in Figure 8.23. The compression is provided by the Materials Testing Systems, Inc. (MTS) testing machine and read by its load cell. Rubber leveling pads compensated for any nonparallelism of the specimen faces. Rotation is permitted about the tapered roller bearing. The specimen is a cylinder coaxial with the MTS testing machine axis. The rubberized cork pads serve two purposes: to dampen the machinery noise and to provide friction for the torque applicator while the compression is maximum. Torque is applied manually by the torque wrench through its torque cell. The AE transducer is held to the specimen by an elastic retainer (rubber band). The procedure for the experiment was as follows:

1. Treat each batch of ceramic cylinders identically.
2. Install the specimen and transducer. Set count to zero.
3. Increase the compression slowly to the maximum and hold until the counting ceased, recording the AE count.
4. Reset count to zero.
5. Increase the torque slowly to the maximum. Hold until the counting ceased. Reduce the torque to zero. Record the count.
6. Analyze the counts for the batch on lognormal probability graph paper.

As it turned out, the data in torque were much more interesting than the results in compression. AE counts in one batch, a typical graph, are plotted in Figure 8.24. The distribution on the graph is the percentage of specimens having fewer than a certain number of counts vs. the number

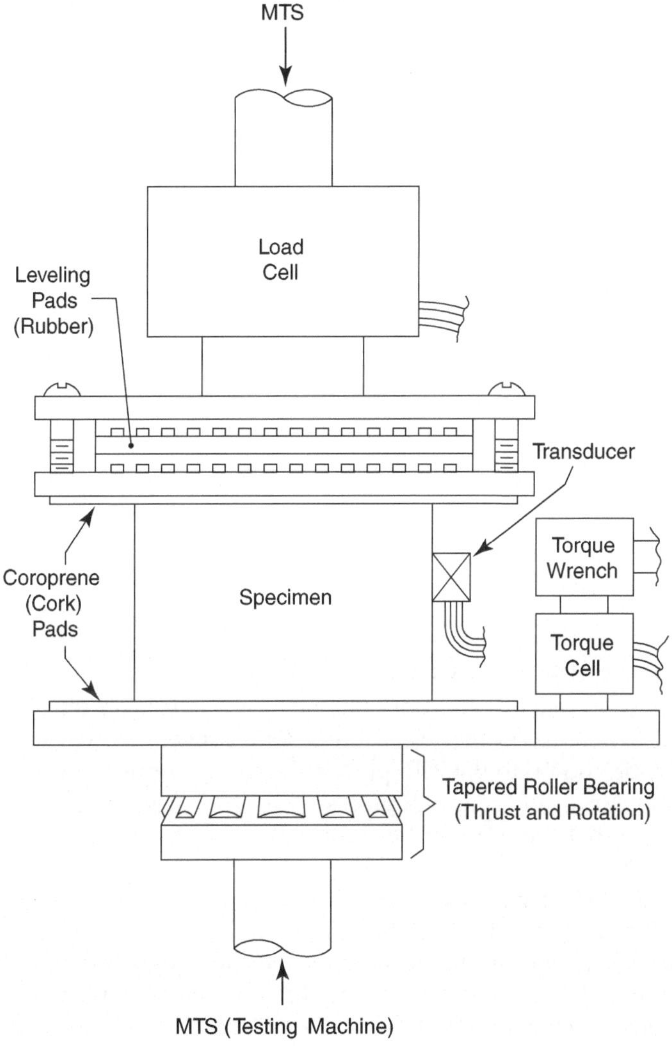

FIGURE 8.23
The fixture in the MTS testing machine. Rotation is permitted about the tapered roller bearing. The specimen is a cylinder. The rubberized cork pads serve two purposes: to dampen the machinery noise and to provide friction for the torque applicator while the compression is at maximum. Torque is applied manually. (From Papadakis, E.P. (1981b). "Empirical Study of Acoustic Emission Statistics from Ceramic Substrates for Catalytic Converters," *Acoustica*, 48(5), 335–338. With permission.)

of counts experienced. The obvious result is that there are two lognormal distributions on the graph. This behavior occurred in all three large batches of different types from different manufacturers tested. It was hypothesized at the time that there was a latent defect of some type in a portion of the specimens in each batch, skewing the results systematically

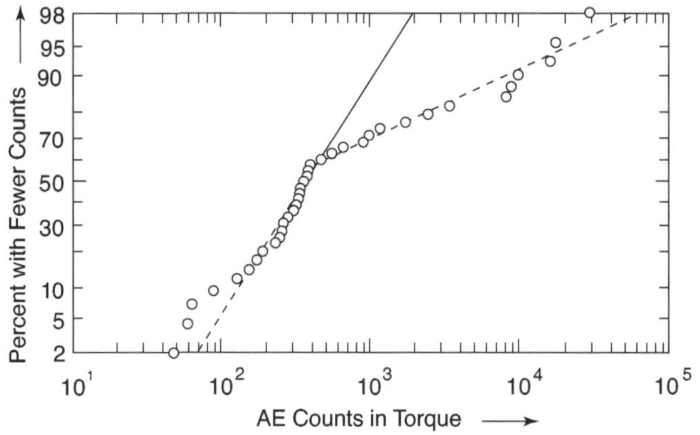

FIGURE 8.24
AE counts in one batch; a typical graph. Two lognormal distributions appear in all three large batches of different types of substrates from different sources tested. A latent defect in a portion of the specimens in each batch was suspected. It was not clear where to set the reject limits. (From Papadakis, E.P. (1981b). "Empirical Study of Acoustic Emission Statistics from Ceramic Substrates for Catalytic Converters," *Acoustica*, 48(5), 335–338. With permission.)

to higher values. That is, it was hypothesized that two distributions were actually detected in each batch.

If one were to use this kind of behavior for quality control, the question arose as to where to set the reject limits. One suggestion was to follow the lower distribution up to the 95th percentile, drop a vertical line to the upper distribution curve, and reject all pieces with counts above this amount. One can visualize that there might be some false accepts and some false rejects because of the statistical nature of the results.

Soon thereafter it was ascertained by road tests of 50 vehicles that the completed and "canned" catalytic converters succeeded in outlasting the governmental regulations of 50,000 miles. An economic decision was made not to complete and roadtest any of the suspected bad parts in the AE distributions. Technical feasibility of the NDT test was not completed because it was determined to be financially unnecessary.

However, the experiment discovered a potentially useful and previously unknown direction for AE to follow in quality assurance of brittle materials. High-strength steels and other alloys, which have limited toughness, can be tested by AE in addition to ceramics. AE can be automated to provide accept and reject signals for tested parts.

8.2.3 Eddy Currents

8.2.3.1 General View of Eddy Currents in NDT

As noted earlier in Section 8.1, eddy currents were discovered almost as soon as transformers for alternating current (AC). While transformers use a magnetically soft iron for their core between two coils, eddy current instruments

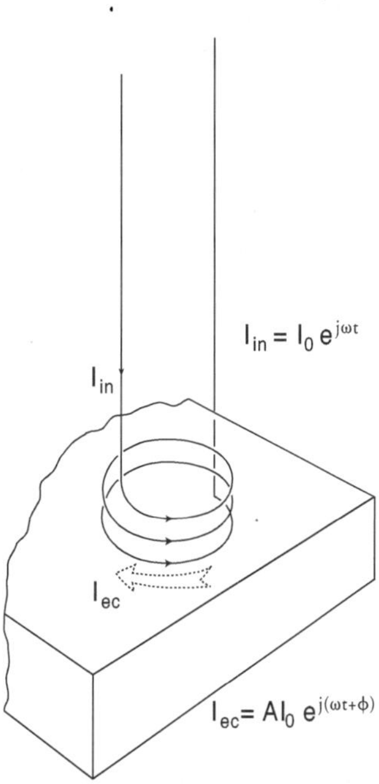

FIGURE 8.25
Production and reception of eddy currents.

use any piece of metal to be tested as if it were the core of the transformer between two coils. Some electrical and magnetic properties of this piece of metal can be deduced, and cracks in it can be detected. It is very important to realize this last pair of facts. Many engineers act as if eddy currents are good only for crack detection. The two types of tests will be treated equally in this book.

8.2.3.2 Production and Reception of Eddy Currents

When a coil carrying an AC of a certain frequency is brought near a metal, eddy currents are generated in the metal in the opposite direction to the current in the coil initially carrying the current (see Figure 8.25). The current in the specimen of metal is induced by the rate of change with time of the magnetic field caused by the current in the coil brought near it. This magnetic field penetrates into the specimen metal only a certain distance given by the skin depth of the metal (see, for instance, Gray, 1957). The eddy currents are induced by the rate of change of this decreasing magnetic field and hence decrease themselves (see Figure 8.26). The skin depth depends upon the conductivity and permeability of the metal and the frequency of the AC carried by the coil. The formula for skin depth is

$$\delta = (2/\omega\mu\sigma)^{1/2} \tag{8.2}$$

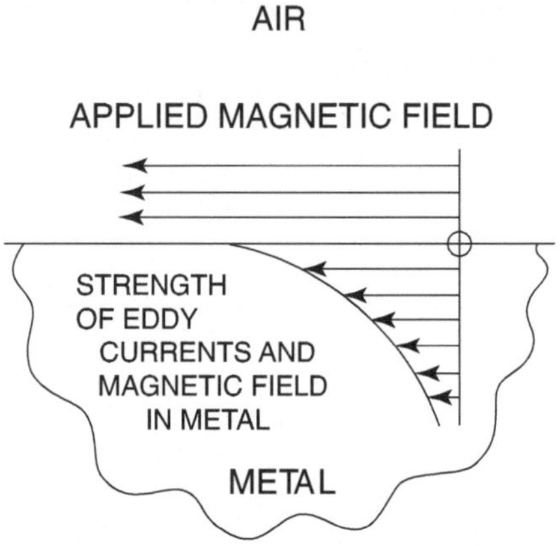

FIGURE 8.26
The magnetic field from the AC in the input coil penetrates into the specimen metal a distance given by the skin depth of the metal. Eddy currents are induced by the rate of change of this decreasing magnetic field and hence decrease themselves. The skin depth depends on the conductivity and permeability of the metal and the frequency of the AC.

where

ω = angular frequency = $2\pi f$,
μ = magnetic permeability,

and

σ = electrical conductivity.

The magnetic field falls off to $1/e$ of its surface strength in the skin depth and continues to decrease exponentially. ($e = 2.71828\ldots$ is the base of natural logarithms.) (Here σ is conductivity, not a standard deviation. Science is running out of Greek letters.) As one can see, higher frequency, higher permeability, and higher conductivity result in shallower skin depth.

The frequency is applied by the eddy current instrument to the coil. There are many designs of coils for special purposes, of course. For the metal, the conductivity and the permeability may be functions of frequency and may be complex quantities such as $A = A' + jA''$. The permeability and the conductivity are changed by the presence of a crack near the coil and by the heat treatment of the metal. Thus, eddy current instruments and coils can be designed to find cracks and monitor metallurgical properties. Due to the skin depth effect, the depth of surface treatments can be measured.

For reception, there are essentially two methods. In one, the impedance of the single coil (as in Figure 8.25) is measured. The induced current in the specimen reacts back upon the input coil, changing its impedance. Because the induced current is a function of both the complex conductivity and the complex permeability of the specimen, the impedance of the coil shows the characteristics of the specimen. For instance, a surface crack will change the conductivity and will change the current in the surface layer of the specimen. In the other method, a receiving coil is used in addition to the transmitting coil. The two may be designed for maximum sensitivity to surface cracks or for maximum sensitivity to metallurgical properties.

8.2.3.3 ***Integrated Instruments and Display Modes***

Many commercial instruments are available. Consult the buyer's guide issue each June of *Materials Evaluation,* the journal of the ASNT. There are three general types as follows.

1. Oscillogram of Impedance Plane. One type utilizes an oscilloscope readout to show the response of the coils in the impedance plane. Several examples of use of the impedance plane are shown in McMasters (1959). The operator is trained to recognize the impedance plane response of the particular coil configuration to the flaw to be detected such as a crack. Some of these instruments are small enough to be hand carried or worn in a chest pack for use in the field where the specimen could be, for instance, an airplane skin.
2. Transient Response/Amplitude Only. The second type simply detects a transient signal from its receiver coil when this signal exceeds a certain threshold. The coils could typically be D-shaped back-to-back in a holder with a small spacing so that if they passed over a crack parallel to the space between them, the transmitted signal would be interrupted (see Figure 8.27). The D-coils could be scanned manually or held in a jig with the parts passed in front of them by automation. This was the type reported in Chapter 4, Section 4.2.6, under Deming Point 6. In that unfortunate case, the coils rotated 90° because of a poor jig design, resulting in a catastrophe. The D-coil design is just fine if it is used correctly.
3. Numerical Components in Impedance Plane. The third type of instrument uses two coils to interrogate a part for intrinsic physical properties. The output current of the second coil is compared with the input current in amplitude and phase in the complex plane. The in-phase and out-of-phase components are displayed electronically or fed into a computer for analysis and recording. Tests for physical properties are designed by constructing conforming and nonconforming samples against which to calibrate the instrument. The instruments can be automated for sorting for quality.

8.2.3.4 ***Specialized Instruments and Applications***

8.2.3.4.1 Gray Iron Hardness

An instrument of the type listed in type 3 above was used to develop a test (Giza and Papadakis, 1979) for gray iron hardness. Gray iron is a flake graphite type for moderate-strength applications. Excess hardness promotes unwanted tool wear in machining operations and inadequate hardness

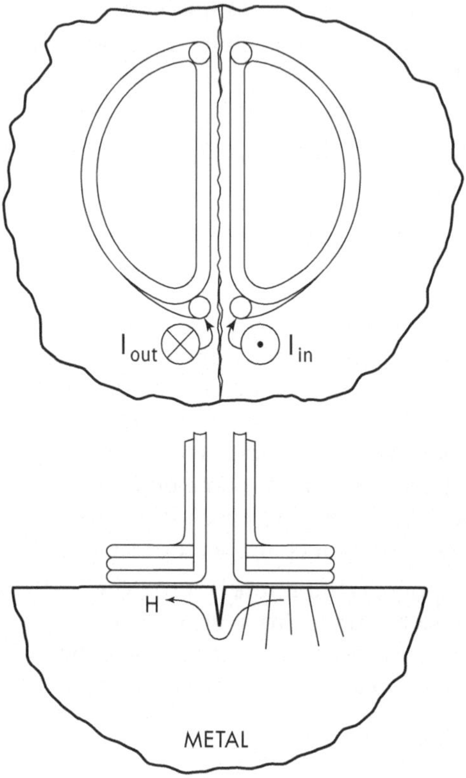

FIGURE 8.27
Coils configured for maximum sensitivity to a crack on the surface between them interrupting the magnetic flux generated by one and detected by the other.

indicated inadequate yield strength. Here hardness is measured by the Brinnell indentation method (Lysaght, 1949).

A library containing a multiplicity of factory-made parts in their as-cast condition was procured and tested with in-phase and out-of-phase components of the output current at various frequencies. The coils surrounded the parts with a high fill factor. The result was that the best correlation was between the in-phase component (A_R in the NDT manufacturer's notation) at a very low frequency (25 Hz) and the Brinnell hardness number (BHN). The resulting correlation is shown in Figure 8.28. As the factory specifications on BHN were 188 to 241, the spread in the data allowed for some false accepts and false rejects when the optimum ECT reject levels were decided upon. This effort was the laboratory feasibility study.

The equipment was moved to the casting plant and the test repeated over a whole week with 600 to 700 samples. The best eddy current reject levels were again determined. The 95% confidence band is shown in Figure 8.29. In the domains around the band and the reject levels, the number of parts

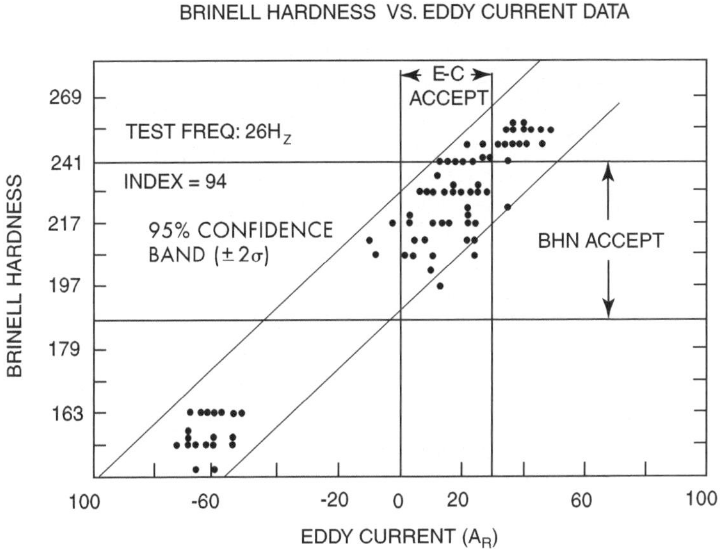

FIGURE 8.28
The best correlation in gray iron was between the in-phase component (A_R in the manufacturer's notation) at a very low frequency (25 Hz) and the Brinell hardness number (BHN). As the factory specifications on BHN were 188 to 241, the spread in the data allowed for some false accepts and false rejects. (From Papadakis, E.P. (1981a). "Challenges and Opportunities for Nondestructive Inspection Technology in the High-Volume Durable Goods Industry," *Materials Evaluation*, 39(2), 122–130. Copyright 1981 © The American Society for Nondestructive Testing, Inc.)

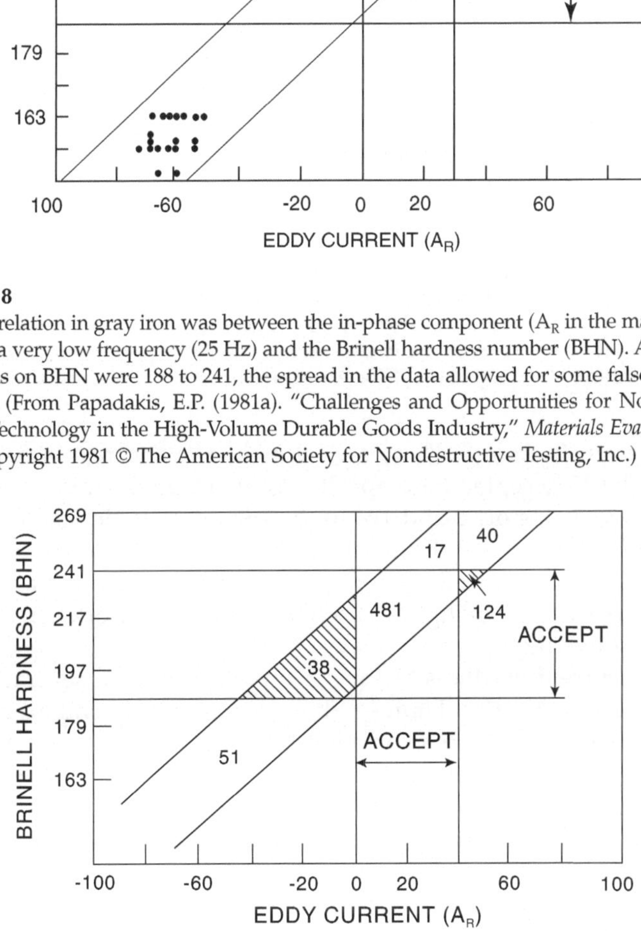

FIGURE 8.29
The 95% confidence band for about 700 samples in the plant feasibility study. In the domains around the band and the reject levels, the number of parts in each domain is shown. There are false accepts and rejects here, too, as expected. The rejected good groups of 38 and 124 in the shaded triangles can be salvaged by performing the regular Brinnell test. (From Papadakis, E.P. (1981a). "Challenges and Opportunities for Nondestructive Inspection Technology in the High-Volume Durable Goods Industry," *Materials Evaluation*, 39(2), 122–130. Copyright 1981 © The American Society for Nondestructive Testing, Inc.)

FIGURE 8.30
An hourly employee fitting a coil over a cast iron part on a conveyor belt to take a reading manually. The eddy current hardness test has been performed on several gray iron parts in several foundries. (From Papadakis, E.P. (1981a). "Challenges and Opportunities for Nondestructive Inspection Technology in the High-Volume Durable Goods Industry," *Materials Evaluation*, 39(2), 122–130. Copyright 1981 © The American Society for Nondestructive Testing, Inc.)

in each domain is shown. There are false accepts and rejects here, too, as expected. The 17 accepted hard specimens are of minimal importance. The rejected good groups of 38 and 124 in the shaded triangles can be salvaged by performing the regular Brinell test. This means that only one third as many Brinell readings need be done as would have been the case without the eddy current test. Savings are accomplished. This work constituted the plant feasibility study.

The eddy current hardness test has been performed on several gray iron parts in several foundries. Figure 8.30 shows an hourly employee fitting a coil over a cast iron part on a conveyor belt to take a reading manually.

One other interesting case is that of parking pawls from the interior of automatic transmissions. These engage a notched ring when the transmission is shifted into park, so the drive train cannot rotate. It is equivalent to a parking brake. If the pawl were to break because of inadequate yield strength, the car could roll into an accident. The yield strength could be inadequate due to inadequate hardness of the pawl. The root cause would be a heat-treating problem. Because yield strength is correlated with hardness, and hardness is correlated with eddy current response, an eddy current test was established for parking pawls. In this case the hardness is on the Rockwell C scale. The defining graph of eddy current response vs. R_C is seen in Figure 8.31. One can see that there are a few false rejects. For this small, cheap part the resultant loss is small. One wants to eliminate all soft parts to avoid accidents. The eddy current test consisted of dropping the parts

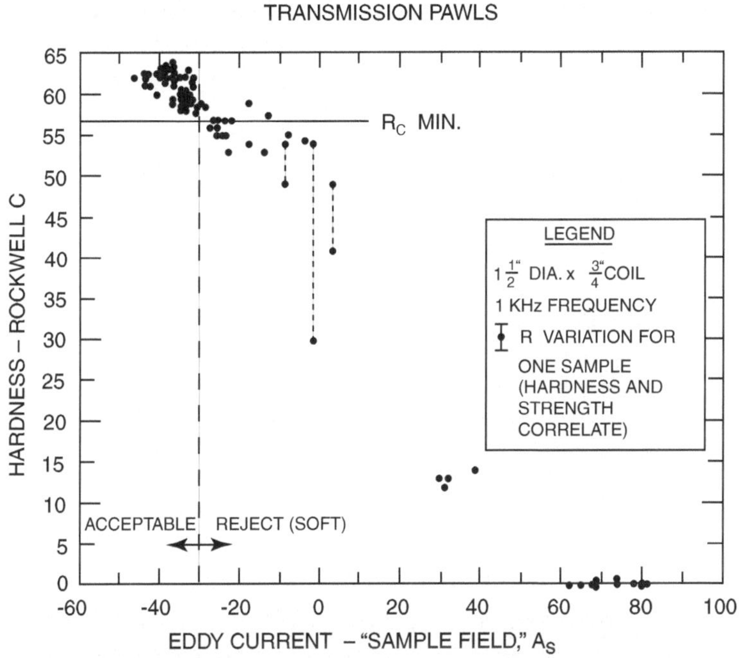

FIGURE 8.31
Eddy current test for parking pawls in an automatic transmission. Pawls of inadequate hardness are to be eliminated. There are a few false rejects but no false accepts.

into a coil and watching for a red light on the instrument panel of the eddy current instrument.

8.2.3.4.2 Case Depth of Steel Axles

Axle shafts are induction heated at the surface and quenched with water spray to produce a hardened case on the tough core steel. The hardened case is necessary to produce high yield strength for the shaft, which experiences flexure and torsion stresses. A hard case is also necessary for bearing surfaces. For quality control, the case depth and hardness must be measured at several locations along the axle.

Using the skin depth of magnetic fields explained earlier in Section 8.2.3.2 to interrogate case depth of hardened regions, an eddy current test was devised (Stephan, 1983; Stephan and Chesney, 1984) for the hardness and case depth of hardened regions along the length of axle shafts for rear-wheel-drive vehicles. An instrument was built to move three coils into position and interrogate them under computer control to find the case depth at six different positions on the axle shaft. Each completed axle consists of two shafts, a right and a left, joining in the differential at the center and mounted in a lubricated structure attached to the automobile. The eddy current measurement was performed on a shaft before assembly.

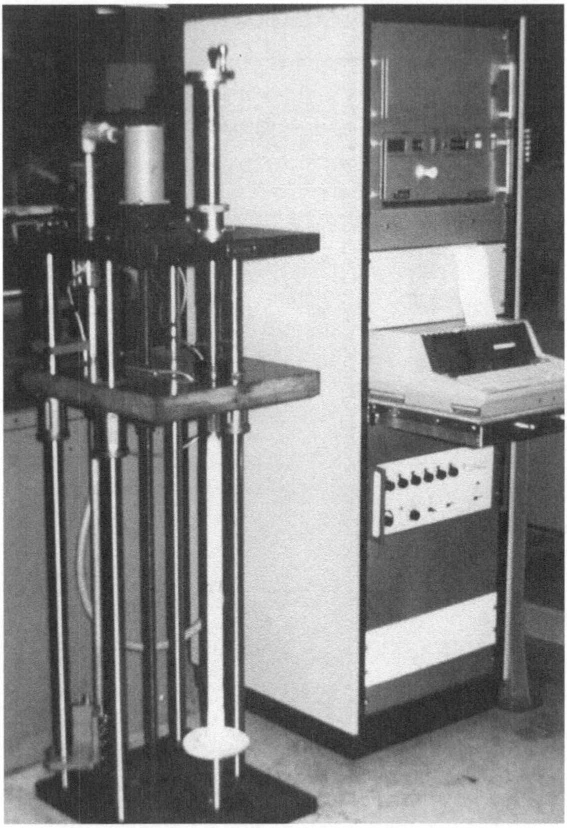

FIGURE 8.32
The completed eddy current system for measuring case depth in axle shafts ready to ship to the axle plant. The computer and the eddy current instrument are in the enclosed instrument rack. The vertical traversing mechanism with the coils and the axle jig is in the foreground. The center square coil carrier on recirculating ball-bearing slides is at the top of the calibration axle shaft, which is painted white. The upper end, which is a spline for fitting into the differential gears, is not painted. The eddy current measurements are made on the retaining knob at the end of the splines (top), on four places along the shaft, and at the curvature where the wheel attachment disc flares out from the shaft.

The completed instrument ready to ship to the axle plant is shown in Figure 8.32. The computer and the eddy current instrument are in the enclosed instrument rack. The traversing mechanism with the coils and the axle jig is the structure in the foreground. The center square coil carrier on recirculating ball bearing slides is a foot below the top of the vertical slide mechanism. The axle shaft, painted white, is held on lathe centers at the front of the traversing mechanism. This axle is the standard for calibration; hence the paint job. The upper end, which is a spline for fitting into the differential gears, is not painted. The disc at the bottom of the shaft will have the five bolts for the wheel press-fitted at a later stage of manufacture.

The eddy current measurements are made on the retaining knob at the end of the splines, on four places along the shaft, and at the curvature where the wheel attachment disc flares out from the shaft. The axle is made by forward extrusion as explained earlier in Section 8.2.1.4.4.2.

The testing system is not fast enough for 100% testing of production, but is used on a sampling basis. It was designed to replace the cutting, polishing, and optical measuring of case depth, which was done at the time the design was conceived. Thus, the eddy current test replaced an expensive and labor-intensive destructive test where the parts cut apart were not inconsequential in cost.

The correlation between the case depth on axles cut, polished, and then measured and the case depth as calculated from the eddy current measurements made by the system is shown in Figure 8.33. This particular graph is for the area along the shaft near step B in the diameter from the extrusion process. Of course, the shaft has been hardened by induction heating and water quenching. A calculated case depth of 100 mils corresponds to real depths between 80 and 115 mils at the 95% confidence limits. Whether this accuracy would suffice would be determined by the chassis engineers to whom the instrument system was to be turned over.

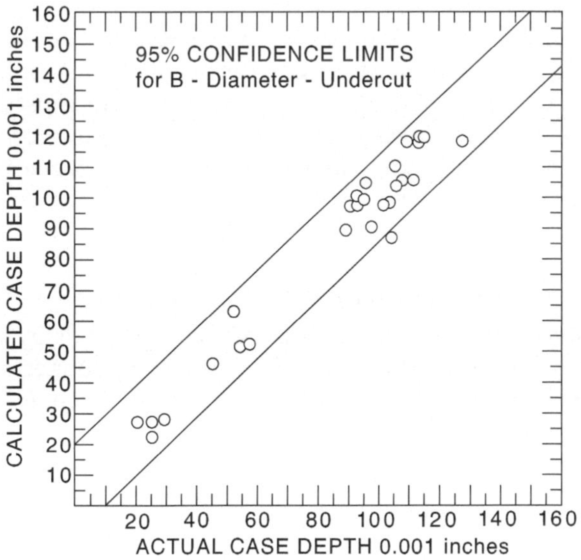

FIGURE 8.33

Ninety-five percent confidence limits for case depth as calculations and measurements are correlated. The measurements ("actual") were made by sectioning and polishing axles. The calculated values came from the eddy current instrument system in Figure 8.32. This correlation is for one location denoted "B" along the axle. (From Papadakis, E.P. (1981a). "Challenges and Opportunities for Nondestructive Inspection Technology in the High-Volume Durable Goods Industry," *Materials Evaluation*, 39(2), 122–130. Copyright 1981 © The American Society for Nondestructive Testing, Inc.)

8.2.4 X-Rays and Fluoroscopy

8.2.4.1 General View of X-Rays

X-rays are penetrating radiation and hence pose a potential danger to health. Factory workers have objected to x-ray tests on this basis, and factory management has often been reticent to risk labor action and even more reticent to spend the money necessary to build the necessary shielding compatible with moving production lines. I once encountered both objections when proposing an x-ray fluorescence leak test for shock absorbers. Another time I had to design a complex sheet-metal shield for an x-ray machine to be used for orienting exceptionally long single crystals of quartz for fabrication into specialized ultrasonic transducers. Technicians and scientists had to work in the same room with the x-ray diffraction machine, not only operating the machine but carrying on other work. Safety was paramount.

The objections are not insuperable when the x-ray methods are necessary and can be designed for human safety. One major example of the use of x-ray inspection is on commercial aircraft D-Checks. The fuselage, essentially stripped to the bare structure, is "wall-papered" with x-ray film on the outside. To be far from people, the aircraft is towed far out onto the apron of the landing field. The film is exposed by portable radiation sources placed along the centerline of the fuselage. One is looking for cracks, especially around windows as stress risers. This is in the realm of maintenance, not manufacturing.

In mass production manufacturing, x-rays can be used in limited situations. The limitation is generally the employees who must interpret the images. For the speed necessary for mass production, fluoroscopy systems are used. To date, to my knowledge, artificial intelligence has not been developed to the degree necessary to eliminate the human inspector.

8.2.4.2 X-Ray Fluoroscopy on Connecting Rods

As a new material, nodular iron cast in permanent molds by an automatic casting process was to substitute for forged steel in connecting rods. The substitution was to be made in an in-line 6-cylinder engine first at one automobile manufacturer. Six rods are required to survive simultaneously in each engine. Stresses are both compressive and tensile.

Because of the complex shape and surface geometry of the near-net-shape castings, it was decided that other scanning methods would not be feasible and that only x-ray fluoroscopic imaging would work as an inspection tool. Fluoroscopy would detect internal voids and possibly cracks. Folds, cold shuts, and external cracks could be seen by visual inspection. It was decided on the basis of cost and availability to have an NDT vendor company do the inspection.

An x-ray fluoroscopic picture of five connecting rods is shown in Figure 8.34. Arrows in the picture point to voids in the cast iron.

In the NDT vendor's equipment, connecting rods on a transparent conveyor belt moved past the x-ray source and its receptor screen. An image of the connecting rod passed across a separate remote viewing screen

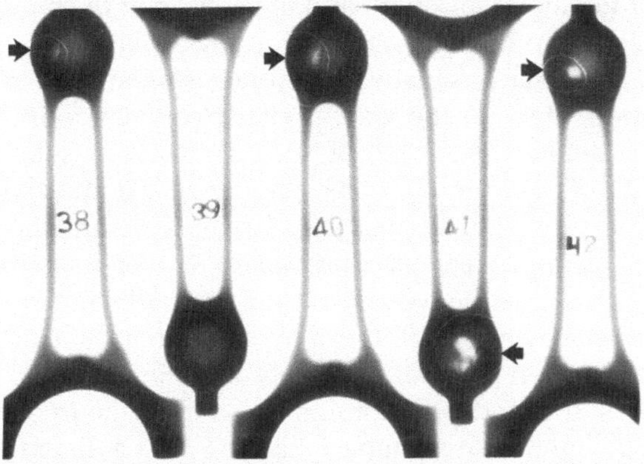

FIGURE 8.34
An x-ray fluoroscopic picture of five connecting rods. Arrows in the picture point to voids in the cast iron. (From Papadakis, E.P. (1981a). "Challenges and Opportunities for Nondestructive Inspection Technology in the High-Volume Durable Goods Industry," *Materials Evaluation*, 39(2), 122–130. With permission. Copyright 1981 © The American Society for Nondestructive Testing, Inc.)

where an inspector was stationed. The inspector made a judgment about the quality of the imaged rod in a matter of seconds and threw a switch for indicating "good" or "bad." The next rod came into view on the moving belt, and the inspector continued. Several systems were running simultaneously.

Two crews of inspectors were required per shift, as the level of attention required meant rest was necessary. Inspectors worked 15 minutes on and 15 minutes off. Accumulated data showed what proportion p of production had voids. The effect of the inspection was quantified. With this arrangement and further visual inspection, no failures came to the notice of the NDT group through the warranty system. The finances of this test (Papadakis, 1985) are analyzed in Section 9.2.4.

8.2.5 Sonic Resonance

8.2.5.1 General View of Sonic Resonance

Sonic resonance is a technique in which a body, when impacted sharply, rings or resonates at characteristic frequencies. The ringing sound is analyzed.

Ordinary bells, tuning forks, leaded glass crystal, fine china, and many other things including some cooking pots ring this way quite noticeably. Things as gross as pilings for architectural structures resonate. One way to drive them into the earth without the ordinary trip hammer on a crane (i.e., a pile driver) is to attach a motor with an eccentric flywheel to the top and run the motor at the lowest longitudinal resonance frequency of the piling.

The piling vibrates along its length and the motion of the bottom end causes the piling to slide into the dirt. On the other hand, things as delicate as the quartz crystal in a quartz wristwatch vibrate similarly in sonic resonance powered by the battery in the watch. That frequency is in the vicinity of 32,000 Hz. The quality factor (*Q*) of such a crystal may be as high as 10 million while the *Q* of a good goblet may be 10,000 and the *Q* of a piling may be 10. The *Q* is the number of vibrations before the amplitude of vibration with no input energy dies out to $1/e$ of its initial value. The constant, *e*, 2.71828…, is the base of natural logarithms.

To use sonic resonance in NDT, the natural vibrations of a body to be tested must continue considerably after an impact. Also, the material property to be investigated must interact with vibrations to change the frequency or the *Q*.

To visualize the motion in an impact, Figure 8.35 is a diagram of the fundamental and the first two overtone modes of resonance of a bar struck on its end. The fundamental is half a wavelength ($\lambda/2$) long. Because $\lambda = v/f$ where f is frequency and v is the mechanical wave velocity (ultrasonic velocity), sonic resonance measures the same intrinsic variables as ultrasonic velocity does. The strain shown in Figure 8.35 is compression and dilatation. The sonic resonance method is most sensitive to properties in the regions of maximum strain, and not sensitive to properties at nodes of strain. In Chapter 9, Figure 9.2 compares the test regions for ultrasound and sonic resonance. Generally speaking, sonic resonance interrogates properties over a much larger volume than does an ultrasonic beam. Some engineers like this averaging approach despite the difficulties with resonance.

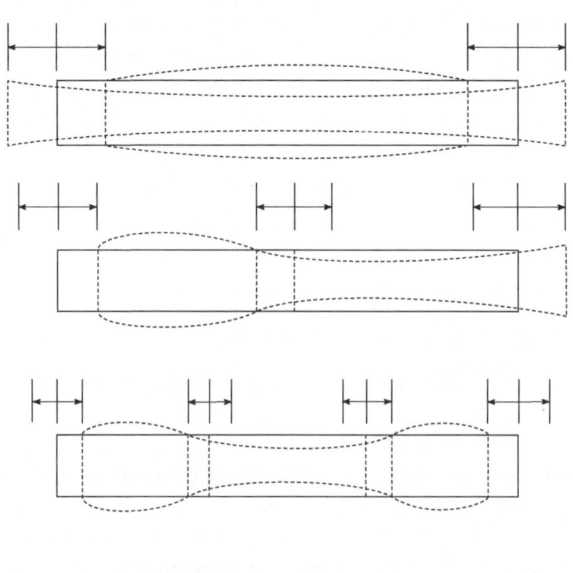

FIGURE 8.35
Diagram of strain, exaggerated, during longitudinal vibration of a rod. It undergoes compression and dilatation. The fundamental and the first two overtones are drawn.

Two difficulties arise from interference due to noise in the environment and from damping (lowering of the *Q*) by the supports needed to hold the piece while impacting and listening. Both these difficulties must be addressed by the design of isolation supports.

8.2.5.2 Sonic Resonance for Automotive Crankshafts

Development was reported (Kovacs et al., 1984) of a sonic resonance system for testing I4 and V8 crankshafts made of nodular cast iron. A specialized frequency and decay analysis instrument already in use for sonic resonance in the automobile industry was procured and adapted for the crankshafts. Initial experiments with the crankshafts supported on rubber chemical corks showed that the fundamental longitudinal resonance would work for a test of even such a complex shape as a crankshaft.

By deliberately casting some crankshafts with improper iron, it was shown that the first criterion could be met, namely that the instrument could distinguish on the basis of resonance frequency between acceptable and nonconforming iron.

The second criterion for a test was to build suitable isolation supports for use in a factory. It was decided that the rubber corks would suffice for the static support at the post end of the crankshaft. That end could be hit by an impactor to generate the sound (vibration). The other end (flange end) also had to be supported on a structure with rubber isolating the crankshaft from the base table. There was a complication as to where to place the accelerometer, which was to be used to pick up the vibration. Attachment directly to the crankshaft was ruled out as impractical for automatic factory operation. It was decided to build a lightweight structure to hold the accelerometer on the crankshaft side of the rubber support under the crank end of the crankshaft. This structure had to move the accelerometer in the same direction as the longitudinal motion of the vibrating crankshaft.

A lightweight structure incorporating rubber Lord mounts® was designed and built. (See Figure 8.36.) The rubber in the Lord mounts supports the weight of the crankshaft while allowing rotation about the center of the rubber. In the configuration as designed, observe the front elevation view in Figure 8.36. The rotation of the rubber permits the left-to-right longitudinal vibration of the crankshaft to be transmitted in the same direction to the accelerometer where the vibration is picked up. The crankshaft rests firmly without sliding on the heads of the two sturdy bolts. The accelerometer output is counted for a given length of time by the instrument to find the frequency of vibration. The entire system is shown diagrammatically in Figure 8.37. Detail of the impactor is shown in Figure 8.38. The instrument and the cradle embodying the designs in Figures 8.36, 8.37, and 8.38 were electronically hard-wired together with a control panel into a testing system.

A laboratory feasibility trial was carried out successfully with both I4 and V8 crankshafts. The system detected the improper metallurgy deleterious to

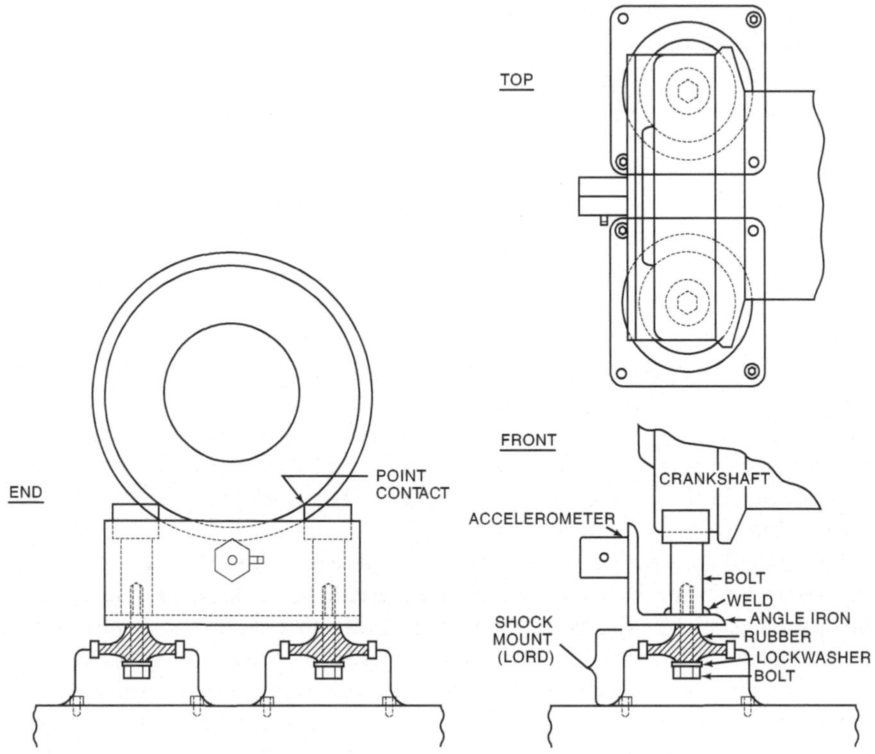

FIGURE 8.36
Three views of the support holding the accelerometer in the test system for the laboratory and plant feasibility trials. The vibration path is from the crankshaft through the 0.5-in bolts and the angle iron to the accelerometer. This structure is isolated from the test cradle by rubber shock mounts that support the weight and permit rotation to pass the vibration. (From Kovacs, B.V., J. Stone, and E.P. Papadakis, (1984). "Development of an Improved Sonic Resonance Inspection System for Nodular Iron Crankshafts," *Materials Evaluation*, 42(7), 906–916. With permission. Copyright 1984 © The American Society for Nondestructive Testing, Inc.)

nodular iron expected through failure modes and effects analyses (FMEAs). It was discovered that parting line flash from the casting process changed the frequency. Note the images of as-cast I4 crankshafts in Figure 8.39. Hence, the test would have to be installed after the shear, which removes the parting line flash before any lathe-turning operations.

The next step was the plant feasibility study. The testing system was moved to a foundry and installed near the parting line shear. Figure 8.40 is a photograph of an hourly employee preparing to load a V8 crankshaft onto the cradle of the test system. The plant feasibility study was successful, and plans were made for factory installations.

The design for the factory installation had the accelerometer fixture near the post end of the crankshaft and the impactor aimed at the flange end. (See Figure 8.41.) A heavy-duty impactor and an accelerometer fixture adapted to the configuration were designed. Details of the accelerometer

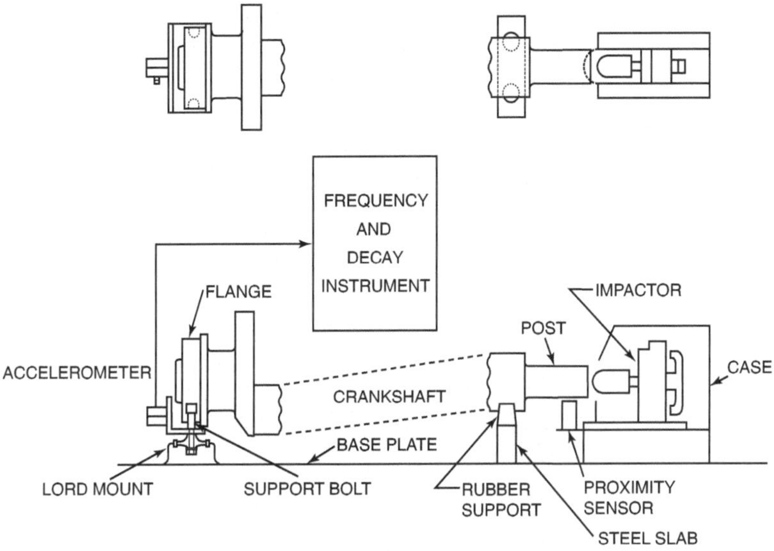

FIGURE 8.37
Test setup for impact on the post end of a crankshaft. This is a diagram of the test system used for laboratory and plant trials. (From Kovacs, B.V., J. Stone, and E.P. Papadakis, (1984). "Development of an Improved Sonic Resonance Inspection System for Nodular Iron Crankshafts," *Materials Evaluation*, 42(7), 906–916. With permission. Copyright 1984 © The American Society for Nondestructive Testing, Inc.)

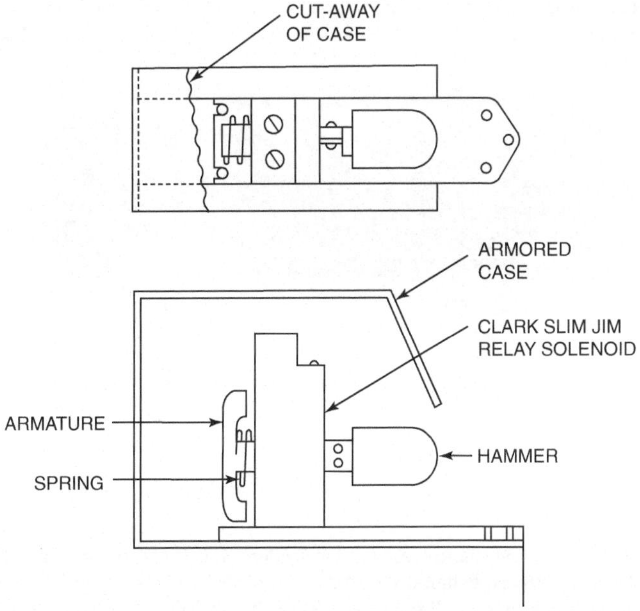

FIGURE 8.38
Details of the impactor in Figure 8.37. (From Kovacs, B.V., J. Stone, and E.P. Papadakis, (1984). "Development of an Improved Sonic Resonance Inspection System for Nodular Iron Crankshafts," *Materials Evaluation*, 42(7), 906–916. With permission. Copyright 1984 © The American Society for Nondestructive Testing, Inc.)

FIGURE 8.39
Photographs showing I4 crankshafts with parting line flash indicated by arrows. (From Kovacs, B.V., J. Stone, and E.P. Papadakis, (1984), "Development of an Improved Sonic Resonance Inspection System for Nodular Iron Crankshafts," *Materials Evaluation*, 42(7), 906–916. With permission. Copyright 1984 © The American Society for Nondestructive Testing, Inc.)

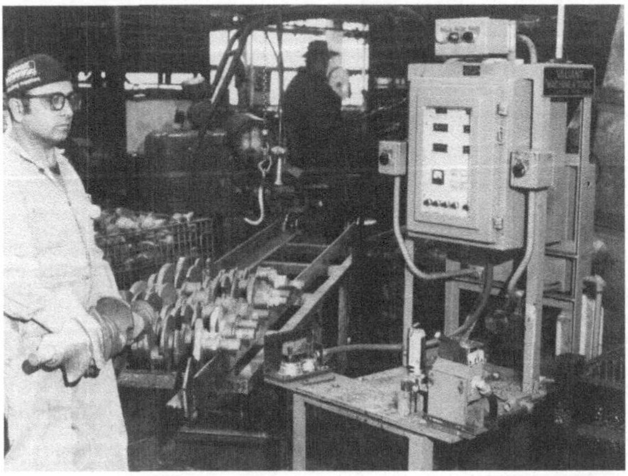

FIGURE 8.40
Photograph of the sonic resonance test system undergoing its plant trial in a foundry. The operator emplaces the V8 crankshaft on the cradle and initiates operation with the two-handed switches beside the control panel. (From Papadakis, E.P. (1981a). "Challenges and Opportunities for Nondestructive Inspection Technology in the High-Volume Durable Goods Industry," *Materials Evaluation*, 39(2), 122–130. With permission. Copyright 1981 © The American Society for Nondestructive Testing, Inc.)

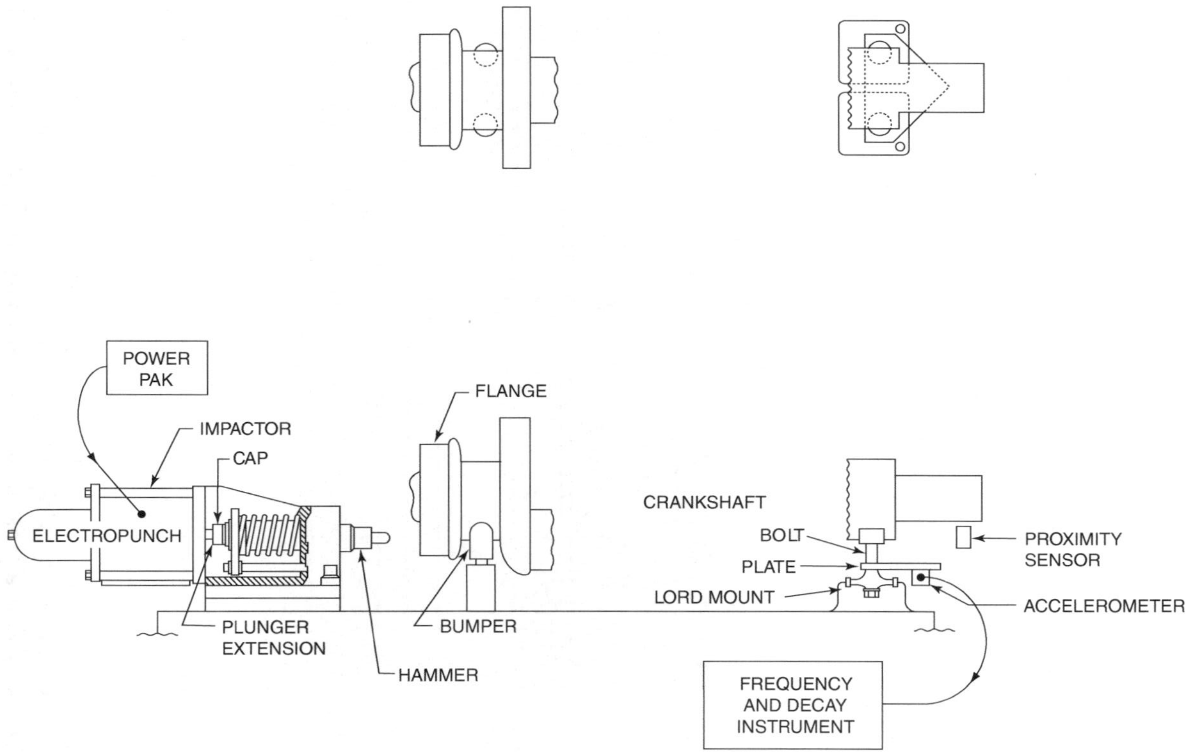

FIGURE 8.41
Test setup with heavy-duty parts for installation in a factory. It is modified for flange impact. The new parts were tested. (From Kovacs, B.V., J. Stone, and E.P. Papadakis, (1984). "Development of an Improved Sonic Resonance Inspection System for Nodular Iron Crankshafts," *Materials Evaluation*, 42(7), 906–916. With permission. Copyright 1984 © The American Society for Nondestructive Testing, Inc.)

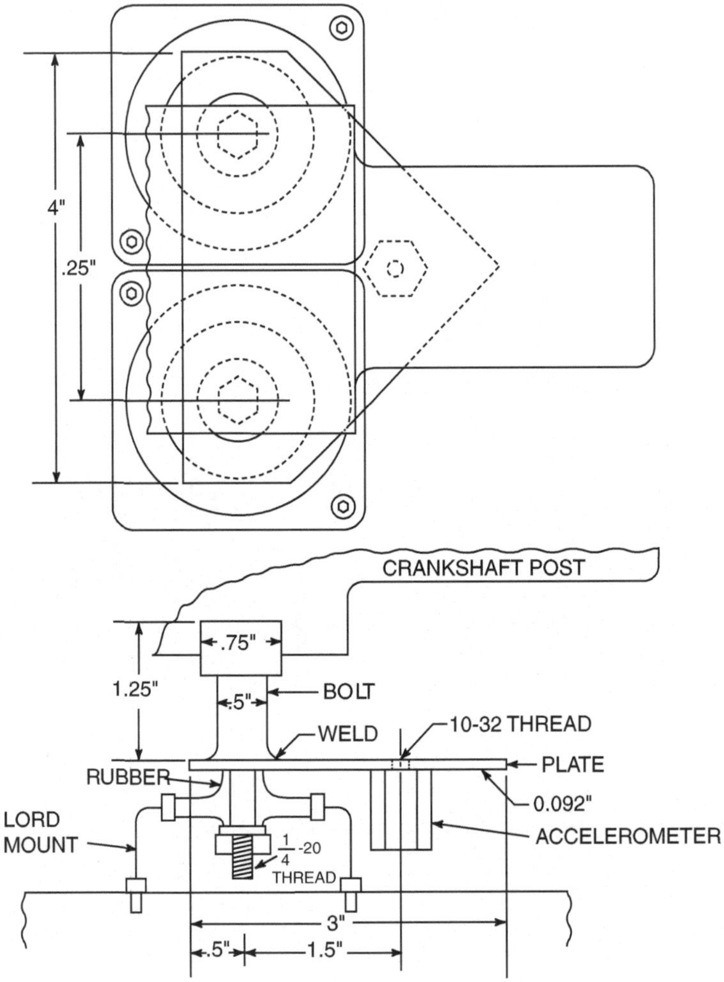

FIGURE 8.42
Details of modified crankshaft support and accelerometer attachment. The principal change is the flat plate instead of the angle iron for the accelerometer attachment. The accelerometer and its cable are offered more protection. (From Kovacs, B.V., J. Stone, and E.P. Papadakis, (1984). "Development of an Improved Sonic Resonance Inspection System for Nodular Iron Crankshafts," *Materials Evaluation*, 42(7), 906–916. With permission. Copyright 1984 © The American Society for Nondestructive Testing, Inc.)

fixture are given in Figure 8.42. The crankshafts were brought to the new cradle by a walking-beam moving transfer line. A crankshaft was set down into the cradle, sensed by a proximity sensor, measured, accepted, or rejected (with spray paint to denote which), and removed. A photograph of the installation is shown in Figure 8.43. The impactor is within the protective grating in the foreground (to keep hands away), and the tubing array is for the spray painting.

FIGURE 8.43
Factory installation of the sonic resonance test system for V8 crankshafts. The heavy-duty impactor is visible in the lower foreground partly covered by a grating hand shield. The crankshafts are transported to and from the test cradle by a walking beam transfer line. (From Papadakis, E.P. (1981a). "Challenges and Opportunities for Nondestructive Inspection Technology in the High-Volume Durable Goods Industry," *Materials Evaluation*, 39(2), 122–130. With permission. Copyright 1981 © The American Society for Nondestructive Testing, Inc.)

As it turned out, there was a time delay while the unwanted vibrations from the walking beam died away, so the transfer line could not be operated as rapidly as the factory management desired. The utility of the instrumentation suffered in the estimation of the management. For later installations, ultrasound was chosen.

Sonic resonance for crankshafts has been treated at length because the development of the method shows the intricacies of bringing a method to implementation. We have gone over initial exploratory work, development, laboratory feasibility, plant feasibility, factory installation, and management interactions. It is believed that the reader will find this project instructional.

8.2.6 Infrared Radiation (IR)

8.2.6.1 General View of Infrared

IR is electromagnetic and lies just below the visible spectrum. That is, the frequency is lower and the wavelength is longer than the corresponding quantities for red light. Infrared is experienced as heat, as for instance while the element of an electric stove is heating up before it begins to glow.

IR radiation can be detected by special photo film and by the photoelectric effect, making various types of "cameras" possible. Germanium can be used for the lenses. The cameras, with continuous viewing and recording capability, vary in size from small camcorders to the equivalent of large TV news cameras carried around on the shoulder and occasionally mistaken for surface-to-air missile launchers. NDT operatives must stay out of the way of SWAT teams.

As for NDT uses of infrared, one must consider situations in which heat is either desired or unwanted. Some interesting and instructive examples are in architecture, transportation, and electric power transmission. A camera aimed at connections in power line wiring can detect overheating in corroding or otherwise bad joints. Inspections can preclude some power failures. Along railroad lines, IR can detect "hot boxes" on railroad axles, a sign of bad bearings and future failure. For buildings, IR can detect heat leaks and, in particular, inadequate insulation. Improvements pinpointed by IR can save heating costs.

Are there useful cases of detecting heat where it is desired?

8.2.6.2 Infrared Assurance of Friction Welds

One useful and instructive case of the use of infrared NDT in manufacturing is on friction welds. A friction weld is made by rubbing two parts together in a reciprocating motion under pressure. Rubbing under pressure generates heat, which finally melts the surface layer of the two parts. The pressure fuses the two together, and then when the motion is stopped, the melt solidifies. The result is a weld. (Rods can be friction-butt-welded by rotary motion without reciprocal motion.)

The parts in this example are two sections of a plastic bumper-reinforcing bar. Bumper-reinforcing bars are the structures that withstand the 5-mph collision or the 2.5-mph collision, whichever standard is to be applied. Early work on infrared monitoring of the friction welds in this part was performed at the Milan, Michigan, plant of the Ford Motor Company circa 1980 using commercial equipment. In the structure in question, one part was a channel beam with rounded edges, and the other was a flat as wide as the outside width of the channel. These two parts, held in jigs, were rubbed together by reciprocating motion of one along the length of the two parts. Force was applied clamping them together to generate the friction during motion to melt the edges of the channel and the extremities of the flat. With the pressure on and the motion turned off, the melted region solidified, making welds along the edges of the channel. This yielded the desired part, a box girder appropriate to be bolted to the front or rear frame of an automobile. The box girder and its attachment means (the PGM tube mentioned elsewhere) takes the impact of a collision but does not show, being covered by decorative plastic fascia.

The manufacturing engineers hoped that the welds actually were made. Good welds all along the box girder were needed for quality. How was this to be ensured? It was decided to use an infrared camera to image the back of the flat surface bonded to the edges of the channel just as the part came out of the friction jig to ensure that the box girder was hot along the back of the two weld lines (see Figure 8.44). A cold area would indicate lack of weld because of lack of melting.

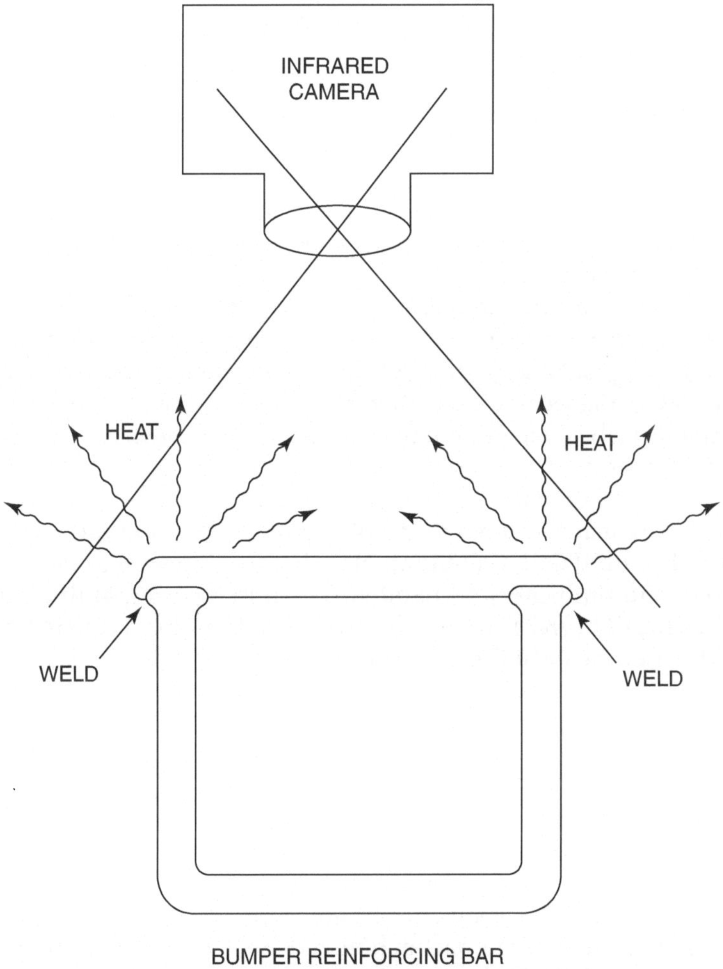

FIGURE 8.44
Diagram of an infrared camera imaging the friction welds in a plastic bumper reinforcing bar just removed from the welding jig.

The camera was installed for a plant trial. When cold areas were detected, the parts were sawed up to confirm the lack of fusion. The method was successful. Some time later, this and further work was reported by G. B. Chapman, who participated in the initial Milan work (Chapman, 2004, 2005b).

The IR method has been automated by using a digital camera feeding into a computer containing an algorithm to detect low values of heat along the supposed weld lines in the bumper case.

8.2.6.3 Other Examples of IR

The author has learned of several other examples of the use of infrared to perform NDT in manufacturing environments. These will be mentioned, but not explained extensively.

Infrared imaging can be used to inspect for various flaws in automotive radiators both for engine cooling and for air conditioning systems (Papadakis et al., 1984). Air blown over the radiator fins interacts with hot or cold fluid pumped through its tubes to transmit or receive heat to and from the air. The face of the radiator facing the air flow is imaged by infrared. Flaws such as clogged tubes, disbonded fins, and the like can be detected as an improper temperature of the fins.

IR can be used as an alternative (Chapman, 2005) to the low-frequency ultrasonic scanner mentioned in a previous chapter (Papadakis, 2002) for adhesive bond quality assurance. A heat source behind an adhesive lap joint will not heat up the second layer where the adhesive is missing or disbonded. The lap joint can be imaged from the unheated side to detect such conditions by low temperature due to poor conductivity.

One of the other parts interrogated by IR was an automotive door structure in which the inner and outer panels were adhesively bonded and partially cured in the stamping plant before shipment to the assembly plant to be put onto cars and painted. The initial curing at the stamping plant was partial (*green state*) and was carried out by inductive heating around the perimeter of the door. At this point the IR imaged the amount of heat applied to the adhesive areas. A large variation was found. Final curing was carried out in the assembly plant by the heat in the paint-curing ovens for the whole car body. It was found that this two-stage curing led to major warranty expenditures due to sagging of the doors. This lack of strength was attributed to inadequate initial cure. The successful IR test was never implemented because of a conflict between the Stamping Division and the Assembly Division over the charge-back of the warranty costs. Chapman (2005b) notes that the automobile company was organized in the compartmentalized or "chimney" fashion advocated by Frederick Taylor so that the financial responsibility was suboptimized, costing the company more than it should have. Reputation was also hurt.

8.2.7 Evanescent Sound Transmission

In the theory of electromagnetic wave transmission from a transmitting antenna to a receiving antenna, there is a region near the transmitter (the *near field*) in which the electric and magnetic fields are extremely complex. Further away, the waves get into the *radiation region* in which the electric and magnetic field vectors are at right angles and are relatively simple expressions. The same complexity occurs in audio and ultrasonic transmission. In the cases of transducers and their fields mentioned earlier, the examples were all in the radiation region many wavelengths away from the transmitting transducer (see, for instance, the discussion preceding Figure 8.5). In the near field, the stress and strain fields are complex. Some energy is trapped in this region and never becomes radiant energy in the far field or radiation region. These waves with trapped energy are termed *evanescent waves*. Their being trapped does not imply that they cannot be detected, however.

Evanescent waves have been put to use in the testing of adhesive lap joints in thin structural materials. Actually, the application uses a mixture of Lamb waves and evanescent waves. One wants to detect and pinpoint the location of small, disbonded regions in the lap joints. The dimension desired is smaller than the wavelength of a convenient Lamb wave for the material. As an example, if the material thickness were 0.1 inch so that two layers bonded with a thin layer of adhesive were possibly 0.22 inches, a convenient wavelength would be 2.0 inches. One wants to detect disbonds smaller than 1.0 inches, so a probe with a transmitter and a receiver 1.0 inches apart would be ideal. This is well inside the near field region of 10 to 20 inches (10λ). Suppose the transmitter and receiver were essentially points. The receiver would pick up some Lamb wave beginning to be transmitted and some energy from the evanescent wave field. The received wave would be different in amplitude and phase for a bonded region and for a disbonded region. Hence, a test for disbonding could be generated.

Indeed, a test just like this has been developed. Rather than describing it here in detail redundantly, the test is described elsewhere. The test is one of several inspection methods used as examples of financial calculations in Chapter 9, Section 9.2.2. The actual technology of testing is described there. A picture of a lap joint is found in Figure 9.3 and a diagram of the probe of the commercial instrument adapted to the test requirements in Figure 9.4. Because of the use of this method to solve a problem relevant to Point 4 in Deming's Fourteen Points, a short report about the solution to the problem was given in Chapter 4, Section 4.2.4. The reader is referred to that section for details of the test method itself and its use. Further published material on the test is given in the several references in Section 9.2.2.

8.3 Correlations and Functions Relating Measurements and Parameters

8.3.1 The Nature of Functions

A function is a relationship stated in mathematics that indicates that a value of x, if known exactly, will result in or predict a value of y exactly. Written out, this is

$$y = f(x) \tag{8.3}$$

The inverse is also true, but may be multivalued. Take, for instance, $y = f(x) = \sin(x)$. In this case, $x = f^{-1}(y)$ is exact but multivalued, the answers being separated by 2. Other sorts of functions like $y = x^2$ yield two values when the inverse is taken, as $x = \pm y^{1/2}$. If the function represents real-world quantities, then the negative answer may be unreasonable.

In the case of real-world quantities such as voltage, resistance, and current, which are known to be functionally related, inevitably there are errors in measuring the quantities. One can conceive of a situation in which the errors turn out to be so large that the functional relationship cannot be ascertained by inspection. Then it is necessary to use regression analysis to fit a curve to the data. The regression may be linear or some curved function. The best curve is chosen by the minimum summed squares of the errors in y away from the curve. This is known as a least-squares fit. Then confidence limits can be computed for this curve. Some curves in Section 8.2 were treated in this way to get the 95% confidence limits.

In a situation with large errors where regression analysis is necessary, the data are approaching the condition of a correlation instead of a function. If one does not know that there should be a real functional relationship but feels that there should be some relation between variables, one may postulate a correlation.

8.3.2 The Nature of Correlations

8.3.2.1 Is There a Relationship?

In correlations, one variable may point to a relationship with another without there being any definitive causative factor. One variable may be predicted from another while there is no cause and effect between them. Often one shows correlations before discovering that there may actually be causative relationships. On the other hand, there may be correlations with no causative relationships at all. In 1956 and 1957, the author saw perfect examples of this situation. The author had the opportunity to participate in a study of floods vs. a number of water-source parameters on tributaries of the Missouri River. Specifically, the maximum instantaneous annual stream flow was being correlated against water source factors such as snow depth on a set of mountains and in a set of valleys. The maximum flow could have been in June, like "the June rise out of the Yellowstone" vs. snow depth accumulations on the February 28 previous. Often, the maximum instantaneous annual stream flow in River R correlated best with the snow cover on Mountain M even though there was no geological or hydrological possibility of water from Mountain M flowing over to reach River R. To predict floods on River R, one would measure snow on Mountain M even though water could not get from there to here. Mountain M did not cause the water in River R, yet the relationship was strong. One had to postulate other variables such as wind patterns and precipitation in January to explain the phenomenon. The research was finally published after much more work by the Missouri River Division of the U.S. Army Corps of Engineers et al. (Missouri Basin Interagency Committee, 1967).

One is led to wonder about the causation in the various medical and nutritional correlations mentioned in the press and published in the best medical journals and health newsletters. If a doctor shows a correlation between the food pyramid and the life expectancy of the Pharaohs, was there any function with causation? Or how about modern carrots and cancer?

The author is not taking a stand on any medical question. However, it is useful to point out, as in the floods-and-snow case, that there need be no functional relationship between the correlated variables.

8.3.2.2 The Need for Relationship

The measurement of the properties of materials is necessary to permit the use of materials in engineering structures. The most fundamental measurements of many properties are destructive. Indeed, the definition of some properties is intrinsically destructive. Examples of this are yield strength and ultimate tensile strength of alloys. The definition of strength involves the pulling of tensile bars cut from representative pieces of the same type of material as will be used in the structure. A useful part can never be tested this way and then used.

Early, primitive methods of testing that circumvented destruction were visual and tactile. Aided by a microscope, one could learn a lot about the properties of a polished metal specimen from its microstructure (ASM, 1985). Properties correlated with microstructure, although an exact prediction was out of reach. Similarly, properties could be correlated with some physical measurements that were not destructive to the part. An example is the correlation of yield strength in steel with indenter hardness (Lysaght, 1949) measurements (Brinell, Rockwell, Vickers). The relationship, of course, resides in the fact that the indenter requires that the material yield to leave the indentation. One can see that if the indentation is on a nonbearing surface in a nonstressed area, the indentation measurement could be considered nondestructive, having no detrimental effect on the serviceability of the part. Thus, the hardness measurement does double duty as a predictor of yield strength. A multitude of specifications have been written on the basis of visual and tactile measurements.

One modern alternative to the labor-intensive methods using trained operators is to automate the old systems. Thus, one arrives at quantitative microscopes and automatic indenter machines, both run by computers that "see" with sensors and calculate as well as control with programmable algorithms. These updated instruments then fulfill the old specifications by measuring in, essentially, the old way.

8.3.2.3 Extending the Relationship

The other modern alternative is to use an entirely different type of nondestructive measurement that also correlates with the property of interest. Rather than having electronics as add-on features, the alternative methods are, generally, intrinsically electronic. Being electronic, they are orders of magnitude faster than the old methods.

Introduction of the new methods suffers from the existence of the old specifications. Frequently, the NDT engineer is required to prove that the new method correlates with the old (accepted) method, rather than with the physical property of interest. This trust in the traditional may be an overriding concern, even when there is reason to believe that the new method is more likely to be functionally related to the property of interest than is the old method.

The present work explores the relationships between a process, its desired output, and the measurable variables also determined by the process. The measurements are studied as correlating with the desired output property and with each other. The possibility of one variable actually being a function of the desired output instead of just displaying a correlation with it is investigated. The conceptual difference between a function and a correlation is discussed below. The effect of interference by extraneous variables is explained as it affects methods such as least-squares curve fitting to find correlation curves. The optimum methodology for arriving at a new specification based entirely on the new method is demonstrated below. Several examples of test development are given.

8.3.3 Theory of Correlations

8.3.3.1 The Underlying Function

A correlation occurs between two variables when there is some intrinsic relationship between the two. The relationship may be causal or only inferential. An example of the latter is found in the prediction of the flow of rivers (Missouri Basin Interagency Committee, 1967) as mentioned in Section

8.3.2.1. The flow of one river may be correlated with the snowfall on a mountain drained by a different river if the two areas receive snowfall from the same weather system. An example of the causal type of relationship was given above between the yield strength of steel and the indenter hardness measurements (Lysaght, 1949). Another example is the correlation between the strength of nodular iron and the ultrasonic velocity in it (Plenard, 1964). This correlation can be related to the shape of the graphite in the iron, which determines the degree of continuity of the iron (strong) as interrupted (Kovacs et al., 1984) by the graphite (weak). This example will be studied further below.

In the causal correlation, the reason that the relationship is not a *function* is that there are third, fourth, fifth, and more, variables involved. A function can be simple like $f = ma$, $V = IR$, $E = mc^2$, or it may be complicated. What distinguishes the relationship as a function is that if the individual variables in the equations are measured to higher and higher degrees of accuracy under the condition that all other variables are held constant, then the data points converge to the theoretical curves to higher and higher precision (Thomas, 1953). With a function, any remaining disagreement can be explained by particular sources of error (Hildebrand, 1956) such as Johnson noise in the resistors and the Heisenberg uncertainty principle for particles.

In the causal correlation, there is an underlying function. However, the extra variables cannot be eliminated, controlled, or measured. The errors relative to the underlying hypothetical function occur on both axes because of the action of the process that produced the thing being measured. (This fact is glossed over in most statistics books and numerical analysis texts where regression is taught. Usually the running variable [x] is treated as error-free while all the error is taken to reside in the dependent variable [see Hildebrand, 1956; Lipson and Sheth, 1973; Martin, 1971.]) Even though one might think that a variable such as ultrasonic velocity could be measured to high accuracy (Papadakis, 1972), say ±1 part in 10^4 or ±1 part in 10^5, this is a measurement on a particular piece, not a measurement on the underlying function.

To further study this concept of the underlying function, consider the fishbone diagram of a process (Scherkenbach, 1986) explained earlier in Figure 3.1. This representation is used in modern TQM to portray all the possible sources of variability in a process. Using brainstorming, the five principal ribs are augmented (as in the diagramming of sentences in grammar) to find all the influences on a process (Scherkenbach, 1986).

Consider each possibility as a variable. Some variables cannot be known. One such variable might be the microcrack distribution (Harris and Lim, 1983) in a piece of high-strength alloy, which might influence the yield strength and the ultrasonic velocity as well as the fatigue life. Similarly, on an even more microscopic scale, the dislocation density, distribution, and pinning (Granato and Lücke, 1956) could influence the yield strength and the ultrasonic velocity. So might chemistry vary when one measurement is made on a coupon from a 50-ton melt. However, hypothesize for the time

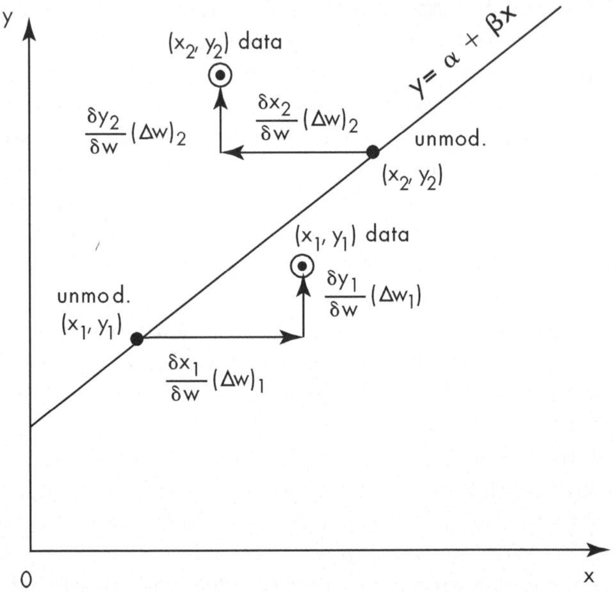

FIGURE 8.45
The underlying function of y vs. x sloping upward to the right is modified by variability in the ordinates and abscissas of its points by the action of another variable, w. The result is a shotgun pattern of points appearing to be a correlation, not a function with errors. (From Papadakis, E. P. (1993). "Correlations and Functions for Determining Nondestructive Tests for Material Properties," *Materials Evaluation*, 51(5),. With permission. Copyright 1993 © The American Society for Nondestructive Testing, Inc 601–606.)

being that *all* the other variables besides x were exactly known and held constant.

Assuming this, the underlying function could be measured as $y = f(x)$. At present consider that this is linear,

$$y = \alpha + \beta x \tag{8.4}$$

This is drawn as a solid line in Figure 8.45. Lift the constraint that the other variables are constants, and consider the effect of variable w upon the point (x, y). The position of x will move by Δx when w changes by Δw, as

$$\Delta x = (\delta x/\delta w)\ \Delta w \tag{8.5}$$

and the position of y similarly,

$$\Delta y = (\delta y/\delta w)\ \Delta w \tag{8.6}$$

The displacement of the points from the underlying function is also drawn in Figure 8.45. When enough points are obtained over a range of x and y values to allow the application of statistics (Hildebrand, 1956; Martin, 1971; and Lipson and Sheth, 1973) (such as least squares) and the drawing of inferences, the resultant set of points will appear spread out as ordinary data in a correlation, which is precisely what the data will be. In addition to w, there will be variables z, u, v, and so on that have added their influences. Inferences drawn from least-squares fits will not refer to the underlying function accurately, however, because ordinary least-squares analysis assumes that measurements on x are accurate representations of the actual points of interest (Hildebrand, 1956; Lipson and Sheth, 1973; Martin, 1971). Instead, one has accurate measurements on displaced points.

8.3.3.2 Origin of Perturbations to the Underlying Function

Consider again the process diagram in Figure 3.1. The process is used in manufacturing and is supposed to result in a desired property. Think of the manufacturing process as casting, heat treatment, shot peening, plating, ion implantation, or whatever. The manufacturing process may be thought of as causing one or more processes in the work piece that produce one or more measurable quantities the work piece can yield up as data. The diagram for three measurables is shown in Figure 8.46. The three processes labeled 1, 2, and 3 may be thought of also as just three aspects of the main manufacturing process (Figure 3.1) which applies the physical determinant at the center of the diagram.

The resultant desired property is in box #1 at the bottom of the diagram in Figure 8.46. It is measured destructively. Box #2 represents a slow or labor-intensive test method, either destructive or nondestructive, which was developed before rapid electronics, and which was found to correlate with the desired physical property. The correlation coefficient is R_{12}. Box #3 depicts a rapid nondestructive test method. This more recent development has a correlation R_{13} with the desired physical property.

However, the NDT engineer may be required by the management to make the NDT method correlate with the old, standard, accepted method rather than with fresh data on the desired physical property. In other words, the NDT engineer may be required to find R_{23}, rather than to go directly to R_{13}.

Mathematically, such a procedure is incorrect; economically, it may be the less expensive course of action in the near term. In terms of mathematics, the NDT engineer is being asked to do the equivalent of finding R_{12} and R_{23} in series. It is well known that the following inequality holds:

$$R_{13} > R_{12} \times R_{23} \tag{8.7}$$

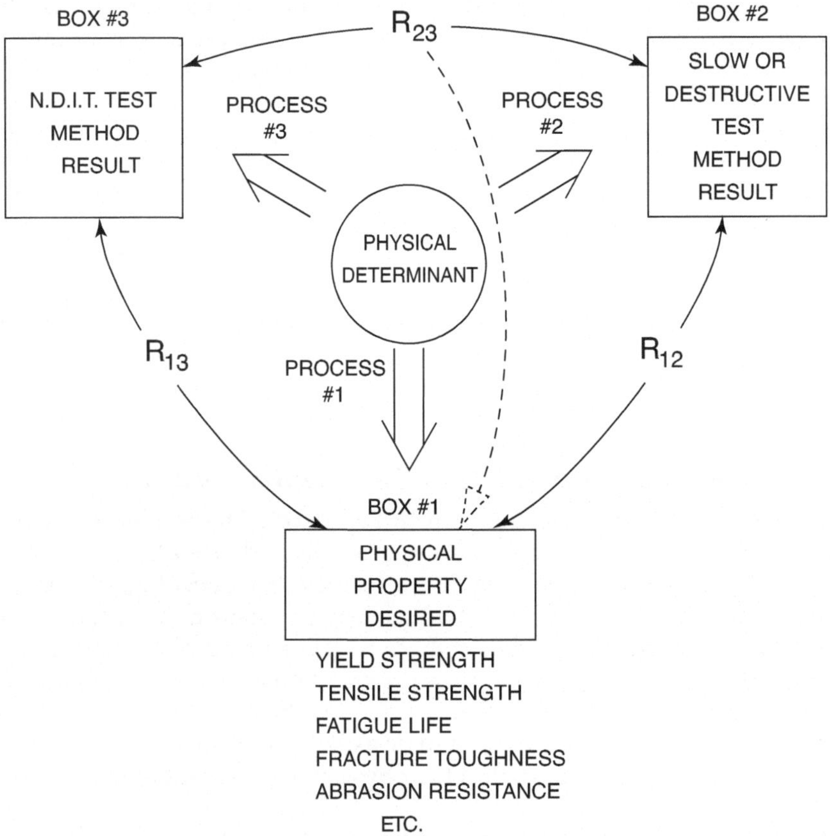

FIGURE 8.46
Model for occurrences within a part treated in a process. Archaic engineering practice may require correlations R_{12} and R_{23} to be used to find correlation R_{13}, whereas it would be better practice to find correlation R_{13} directly. (From Papadakis, E.P. (1980). "Correlations and Functions for Determining Nondestructive Tests for Material Properties," *Materials Evaluation*, 51(5), 601–606. With permission. Copyright 1993 © The American Society for Nondestructive Testing, Inc.)

at all times when $0 < R_{12} < 1$ and $0 < R_{23} < 1$. In all cases of interest, the correlation coefficients are indeed not perfect (i.e., below 1.0).

The inequality in Equation 8.6 can be visualized by solid geometry. The data sets being correlated can be thought of as vectors emanating from a point and directed into a single octant of space. The correlation coefficients between pairs are equal to the cosines of the angles α, β, and γ between the vectors. Because no two angles can sum to less than the third angle, the inequality

$$\cos\alpha > \cos\beta \cos\gamma \tag{8.8}$$

always holds for any permutation of the angles. (Try it on a pocket calculator.) Hence, finding R_{23} when R_{12} is smaller than 1.0 and greater than zero is not as good a way of establishing the validity of the NDT test, which would be finding R_{13} directly.

The process diagram in Figure 3.1 indicates that there may be a multiplicity of physical determinants, like the one at the center of the diagram in Figure 8.46, operating on the part in question. These other operators interject the extra variables *w*, *u*, *v*, and so on. They might be preexisting conditions, too, as in the raw materials. It would require more than three dimensions in space to diagram all the possibilities. The best correlation will always be between the two variables actually desired. In our case, this is the physical property desired and the NDT measurement.

8.3.4 Experiments with Correlations

Two experiments have already been described in which correlations were found between a physical quantity of interest and an NDT parameter. These will be summarized here.

An eddy current test was described in Section 8.2.3.4.1. The reader is referred to that section and its figures for the details. The important point to note is that a correlation was required by management between the eddy current reading and the indentation hardness reading in the type of iron in question. What was desired in actuality was the yield strength of the iron. It was known (Lysaght, 1949) that the indentation hardness correlated with the yield strength. Specifications had been written for the iron in terms of the indentation readings. Because of the reliance upon the old, traditional method, the NDT engineers were required to do serial correlations from eddy current readings to indentation readings to the final result—strength. The result was suboptimal but still useful. Refer back to the text referring to Figure 8.29, where reject limits for use in the eddy current test are discussed.

An ultrasonic velocity test was described in Section 8.2.1.4.2. The reader is referred to that section and its figures for the details. The important point to note is that a correlation was developed between the ultrasonic velocity and the physical quantity actually desired—the yield strength—by making the ultrasonic velocity measurements on iron, from which tensile bars were also made. In pulling the bars, the ultimate tensile strength was also found and a correlation established with ultrasonic velocity for that variable in addition. The old, traditional specification of optically read nodularity was bypassed to arrive at an optimal correlation. As one will note in Figure 8.14 shown earlier, the correlation is tight enough so that the curvature of the nonlinear underlying function can be observed. The yield strength curve provides reject limits for the ultrasonic test.

8.3.5 Generic Curve for Reject Limits

In the examples of test design above, the observation was made that some acceptable material might be rejected and that some nonconforming material might be accepted by an NDT using correlations or regressions with error bands. In using NDT for quality assurance (QA), it is inevitable that such errors occur. In the parlance of the quality profession, they are termed Type I errors (calling good material bad, or false rejects) and Type II errors (calling bad material good, or false accepts). The probability of detection, Figure 8.12, is related to this conceptually.

Such errors also occur in statistical quality control when a few nonconforming parts elude the sampling process (Enell, 1954). However, with 100% NDT (applying NDT), the effect of the same percentage of Type II errors is much more benign (Papadakis, 1982) because only parts slightly out of specification can be outside the confidence limits in the vicinity of the reject set point. This fail-safe feature is shown in Figure 8.47. This figure may be considered a generic picture of reject limits in any case in which there are data of a desired design parameter vs. an NDT parameter. For convenience, Figure 8.47 is drawn as if there is a linear correlation where 95% confidence limits have been calculated to go

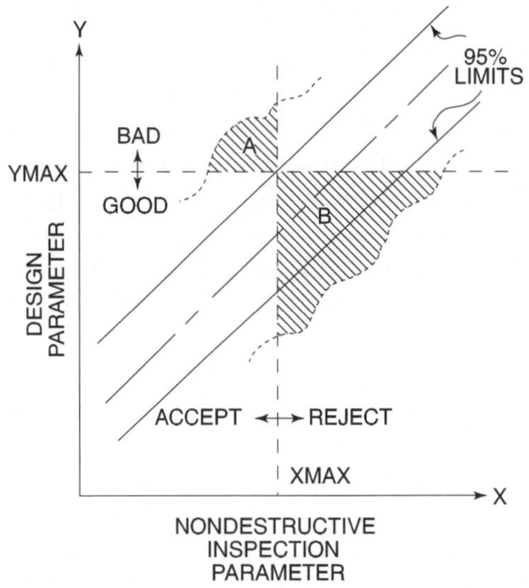

FIGURE 8.47
Diagram of accept–reject levels of an NDT test relative to the acceptable vs. nonconforming level of performance of a material. The case of a positive correlation slope with a maximum permissible value of Y, namely YMAX. The positive slope with YMAX determines that there will be a maximum allowable value XMAX for the NDT parameter. The shaded area A represents false accepts or Type II errors.

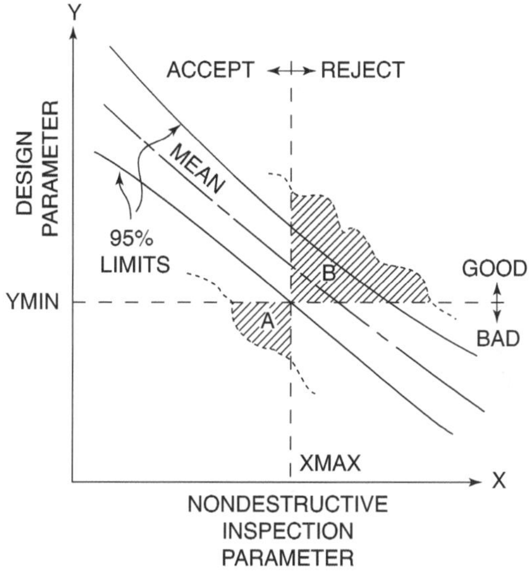

FIGURE 8.48
The case of a negative correlation slope. With a YMIN specified in this case, the negative slope determines that there will be a maximum allowable value XMAX for the NDT parameter. The shaded area A again represents false accepts or Type II errors.

with the regression line. The design parameter must be no higher than YMAX, so the NDT reject limit, XMAX, is found from the intersection of the upper 95% confidence limit and the YMAX line. As a few parts will be outside the 95% confidence limits, there will be a few Type II errors (false accepts) in the small shaded area, A. These are "bad" by a slight amount, meaning that they will tend to be benign compared with a part missed at random in sampling.

Other combinations can be calculated if one has a negative slope or if one has a YMIN instead of a YMAX to contend with. Occasionally one may have a range of acceptable values of Y so that there are both a YMIN and a YMAX.

The case of a negative slope is shown in Figure 8.48. With a YMIN specified in this case, the negative slope determines that there will be a maximum allowable value XMAX for the NDT parameter. The shaded area A again represents false accepts or Type II errors.

The case of a positive slope with a permissible range of the design parameter between YMIN and YMAX is shown in Figure 8.49. One finds an acceptable range of the NDT parameter between XMIN and XMAX. There are two shaded regions of false accepts, A_1 and A_2.

The case of a negative slope with a permissible range of the design parameter between YMIN and YMAX is shown in Figure 8.50. One finds an

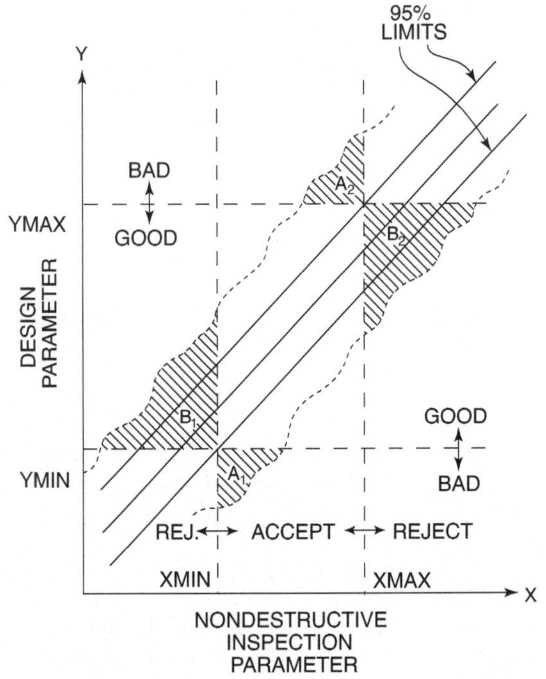

FIGURE 8.49
The case of a positive correlation slope with a permissible range of the design parameter between YMIN and YMAX. One finds an acceptable range of the NDT parameter between XMIN and XMAX. There are two shaded regions of false accepts, A_1 and A_2.

acceptable range of the NDT parameter between XMIN and XMAX. Again there are two shaded regions of false accepts, A_1 and A_2.

8.3.6 Summary of the Correlation Approach

When a process results in three measurable outputs where 1 is the property of interest, 2 is a property used in specifications to test for property 1, and 3 is a property proposed to supplant 2, the proper methodology to use is to perform a correlation R_{13} between 1 and 3 directly. Poorer results will be obtained by insisting upon finding R_{23}. Equation 8.6 states this fact categorically.

8.3.7 Philosophy of the Scientist and the Engineer

The difference between functions and correlations is mirrored in the outlook of scientists and engineers. In attempting to establish a test method, the

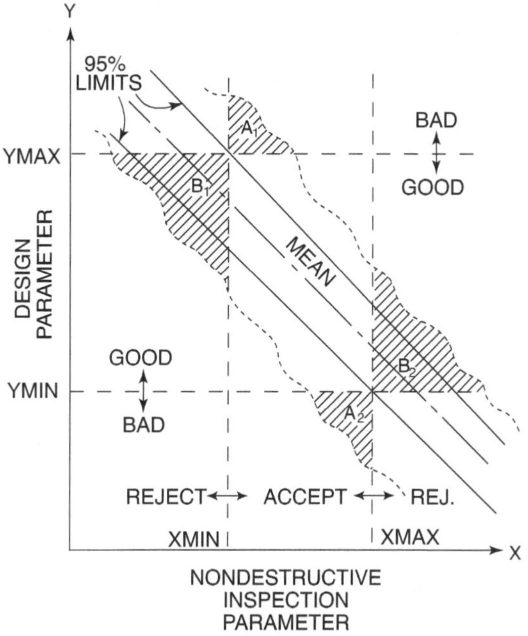

FIGURE 8.50
The case of a negative correlation slope with a permissible range of the design parameter between YMIN and YMAX. One finds an acceptable range of the NDT parameter between XMIN and XMAX. Again, there are two shaded regions of false accepts, A_1 and A_2.

scientist and the engineer will both undertake experiments. However, the character of their experiments will differ.

The scientist will set up an experiment with one dependent variable and one independent variable. In this experiment, all other possible variables will be held constant. A minimal number of specimens to cover the range of the dependent variable will be obtained from a single (hopefully invariant) source, and measured. Typically there would be four to six specimens covering the range to define the functional dependence of the dependent variable upon the independent variable. If the results are a smooth curve, the scientist will be satisfied and will infer a law from the curvature (or linearity). Should the data exhibit a lack of smoothness, the scientist would tend to obtain three more specimens at each of the (say) six points along the range axis so that the amount of error at each point could be ascertained. The textbooks say that one specimen can be thrown out among each four if it deviates more than three standard deviations from the mean of the set of four. Thus, the scientist would hope to throw out a few points that had caused the original distribution not to be smooth, and to place error bars upon the saved points. A function or law with errors would be the output.

As a test, the result would have a 95% confidence limit with the constraint "all other things being equal."

The engineer comes to an experiment from a different background in both education and practical experience. The engineer will be keenly interested in the possibility of generating a valid test in the presence of all the variability allowed within the specifications of the process portrayed in Figure 3.1. The process, while under control within these specifications, will make both acceptable and nonconforming parts (Western Electric Co., 1956). An out-of-control process will result in many more nonconforming parts. The engineer is principally interested in eliminating all nonconforming parts, within the context of the allowed input variability and the possible deviations. To do this, the engineer will approach the problem with two complementary types of action: (1) Obtain a multiplicity of work pieces made by the process over a time period long enough to represent most of the possible permissible input variability. (This may involve waiting for parts from several batches of material from all suppliers, for instance.) (2) As in design of experiments (Lipson and Sheth, 1973), the engineer may make process changes outside the limits of the specifications to produce nonconforming parts deliberately. The input variability on other variables will be maintained while changing the chosen variable (several will be chosen). The variations to be made will be influenced by failure mode and effects analyses (Ford Motor Co., 1979). It must be emphasized that action 1 is radically different from action 2. Action 1 is not *design of experiments* in the current definition of the term (Lipson and Sheth, 1973).

The entire library of specimens will be measured by the proposed NDT method and then tested by the fundamental method defining the property of interest. The data will be treated as a correlation of the values of the property of interest vs. the NDT data.

This approach was taken in the two examples (Giza and Papadakis, 1979; Papadakis, 1976b) shown in the section on experiments with respect to procuring the libraries of specimens. One deviation constrained by management was the correlation against BHN instead of the property of interest in one case.

For the development of a valid NDT test, the process given above for amassing a library of specimens should be followed.

8.3.8 Conclusions Concerning Correlations

The correlation approach to data analysis takes into account the shift of data points in both the x and y directions away from an underlying function $y = f(x)$. The shift of the data points is caused by the action of uncontrolled variables that are intrinsic to the manufacturing process being carried out. Even while the process is under control, these extra variables produce some shifting of the data points. When the existence of these variables is recognized, a proper experiment can be devised to take their variance into account to produce a correlation for prediction purposes. Such a correlation is desired for quality assurance. It is desirable that the method of measurement be nondestructive and rapid. While it is best to produce a direct correlation between the NDT test and the property of interest, it is also possible but less accurate to set up a correlation between the rapid NDT test and some other test that has become a traditional, specified test method. Examples have been given showing the use of the correlation method with ultrasonic and eddy current tests.

9

Real Manufacturing Examples of the Three Financial Methods of Calculation and of Real Decisions Made on the Basis of Those Calculations

9.1 General

Before the nondestructive testing (NDT) expert is called in, the production engineering staff will have a general idea of the proportion of nonconforming parts being produced in the process to be tested. This knowledge may arise from warranty feedback from dealers, from batch testing of outgoing product, or from some other indication of a breakdown of the system. If the product or process is new, the proportion of nonconforming product may be inferred from previous experience or may be predicted by a failure modes and effects analysis (FMEA). The team doing continuous improvement may report that the present status of the process is a certain proportion of nonconforming parts it hopes to reduce to a lower figure in a particular length of time. Generally, a management judgment will be made that the present level of nonconformities is unacceptable. Then NDT is called in to create a fix for the duration. NDT will be used if it is cost-effective in the sense of one of these three methods. The duration may be until continuous improvement reduces the problem to a level acceptable to management, until the problem is reduced to a point where the NDT is no longer cost-effective, until the part is phased out, or for an indeterminate time far into the future. In any case, the NDT is to be used for 100% inspection of production for the duration.

For the calculations in the three methods, the cost data for Sections 6.2.2 and 6.2.3, as well as all other cost and production data, must be current. In the examples cited in this chapter, the data were current for the time period in which the inspection decisions were made. The data may seem old, for instance using 1988 economics to choose to test or not to test in 1988. It may be that costs for the same failure in 2004 might be higher. One must study his own applications on a case-by-case basis. The rate of increase in detrimental costs may not equal the rate of increase in testing costs. It is possible, for instance, that the testing instrumentation may become cheaper along a learning curve, while the cost of the product to be tested increases with inflation. The cost of capital (i.e., the interest paid on borrowed money to buy equipment) may vary greatly over decades. Within memory the Federal Reserve has set interest rates as low as 1% and as high as double-digits. Along with the cost of money, the psychology of inflation may impact different businesses differently, and the hurdle rate the controller may quote for buying equipment may be sky-high during double-digit inflation. Regardless, the costs are to be calculated at current and projected economics. Costs in the quoted examples were for the years in which the failures occurred.

9.2 Examples of the Deming Inspection Criterion (DIC) Method

These examples were first presented in "The Deming Inspection Criterion" (Papadakis, 1985). In that reference, the examples were much abbreviated and condensed to be relevant and yet generic.

9.2.1 A Process with Each Part Unique: Instant Nodular Iron

This process is initially described generically to demonstrate the breadth of applicability of the inspection concept. When each part created is unique, there is no way to do batch traceability or to test only a few out of a definite larger group. The inspection must be performed on all parts because each one could fail independently of any other.

By contrast, in some chemical processes, it is possible to make a large mixture, use it to make parts until it is exhausted, and then be assured that all the parts are good if the last part is good. In other words, the quality control (QC) test could be performed on only the last part made. If it is good, then all ahead of it would be good. Although 100% inspection would not be needed, batch traceability would be required.

This is true of cast nodular iron made by mixing a given amount of magnesium ferrosilicon into a ladle of molten iron and then pouring it into a series of molds in a timely fashion. This mixing is termed *inoculation*. The function of the magnesium is to cause the carbon in the molten iron to grow into microscopic spheres throughout the solidifying iron. If the graphite is in spherical shape, then the whole casting has maximum strength because the iron is contiguous around the graphite to the maximum degree (Papadakis et al., 1984). The effect of the magnesium fades over time so that *if* the pouring is not done soon enough, some of the parts in the later molds will not be good. Testing the last part poured will show whether all the parts are good or if many more must be tested to find out how far back in time the fading became too aggravated.

In the new process being introduced, the ladle is not treated with the additive. Rather, small amounts of the additive, as solid lumps or granules, are put into the runner system of the molds so that each mold gets a bit of additive, which dissolves in the molten iron as it is poured into the runner. Thus, each mold is unique, having its own source of dissolving additive. The process is called *in-mold inoculation*. Up to 200 molds may be poured from a ladle. There are various failure modes, such as oxidation of the additive slowing its dissolving, overly rapid dissolving of the particles, not putting enough additive into the runner, inadequate mixing of the solute into the solvent, and so on. Magnesium-poor regions can form, weakening an otherwise strong part. Figure 9.1 is a photomicrograph showing iron with

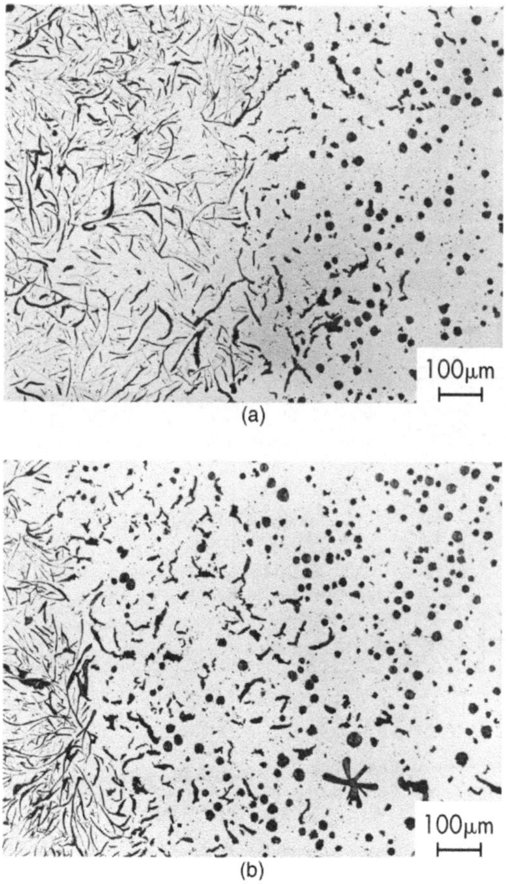

FIGURE 9.1
Photomicrographs of iron made by the in-mold inoculation process with inadequate magnesium in one part, resulting in flake graphite in part of the volume. (From Kovacs, B. V., Stone, J., and Papadakis, E. P. (1984). "Development of an Improved Sonic Resonance Inspection System for Nodularity in Crankshafts," *Materials Evaluation*, 42(7), 906–916. With permission. Copyright 1984 © The American Society for Nondestructive Testing, Inc.)

adjacent nodular and gray iron areas. The graphite nodules are close to spheres, while the gray iron is characterized by flakes of graphite that show up edge-on upon the polished surface. This piece of iron was made by the in-mold process with inadequate magnesium in the gray iron area.

Production cost and reduction of pollution (boiling off magnesium from ladles) were two of the major drivers motivating the introduction of the in-mold inoculation process. The parts to receive the new process were I4 (in-line four-cylinder engine) crankshafts. The NDT development group in the company was called in to produce a method of testing if feasible and cost-effective. The group knew that ultrasonic velocity would provide a feasible test at predictable costs. Figure 9.2 schematically depicts ultrasonic beams

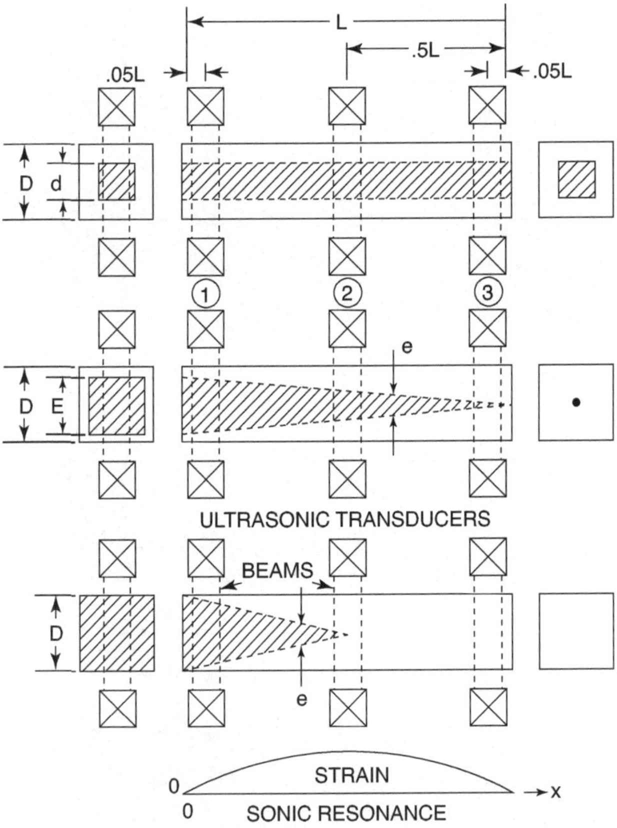

FIGURE 9.2
Ultrasonic beams traveling from one probe to another through rod-shaped parts representative of crankshafts. The shaded areas represent poor iron with lower ultrasonic velocity. These deficient areas would have resulted from lack of magnesium as too little was available at one end of the casting. The ultrasonic beams would have picked up the lower velocity. (From Papadakis, E. P. and Kovacs, B. V. (1980). "Theoretical Model for Comparison of Sonic-Resonance and Ultrasonic-Velocity Techniques for Assuring Quality in Instant Nodular Iron Parts," *Materials Evaluation*, 38(6), 25–30. With permission. Copyright 1980 © The American Society for Nondestructive Testing, Inc.)

between pairs of probes going through rod-shaped cast parts representative of crankshafts. The shaded areas represent poor iron with lower ultrasonic velocity. These deficient areas would have resulted from lack of magnesium, as too little was available at one end of the casting. The ultrasonic beams would have picked up the lower velocity. Costs were the question because the crankshafts were not considered a critical safety part. Failure of a crankshaft would be very bad publicity for the company and would entail an expensive warranty repair.

Experience and FMEAs predicted that a proportion p of nonconforming parts of 1/5000 = 0.0002 or greater was to be expected. The detrimental cost of replacement or repair was estimated as \$1000 on average. The cost of testing a single part, k_1, was calculated to be \$0.20. The latter included the cost of a commercially available instrument expensed in the first year, and the cost of an operator on the plant floor working at a reasonable rate. Added to the repair cost was a figure of \$1000 for deleterious effects upon reputation impinging upon future sales, making the resultant k_2 to be \$2000. Using Equation 7.1, the result is

$$\begin{aligned}\text{DIC} &= (k_2/k_1) \times p \\ &= (\$2000/\$0.20) \times (0.0002)\end{aligned}$$

or greater, so

$$\text{DIC} \geq 2.0 \tag{9.1}$$

which indicates that 100% inspection should be initiated. The factory did, indeed, institute 100% inspection. Equation 9.1 shows that the inspection should continue until metallurgical improvements in the in-mold inoculation process might bring the proportion of nonconforming parts below 1/10,000.

9.2.2 Adhesively Bonded Truck Hoods: Sheet Molding Compound-Type Fiber-Reinforced Plastic (FRP)

A new method of joining materials permitted the introduction of a new material into a major subassembly. Other subassemblies and parts using the method and material were planned. However, the initial quality assurance methodology for the joining method was found to be inadequate after an expensive problem developed. Both the product and the future plans were in jeopardy.

This generic description can be interpreted by the reader to cover his or her own set of problems. However, the specific problem is expanded upon here.

The product was the skins for truck bodies. In particular, the product of immediate concern was heavy truck hoods. The hood was made of two sides, a right and a left. Each side connected the fender and the hood to the cab section in one complex part. Openings for the grill, headlights, and air cleaner intake were left in the sides where necessary. The sides were molded in heated presses. Layers of the sheet molding compound (SMC), a heat-setting plastic with about 30% chopped glass fiber about 2 inches long in random directions, were laid into the female side of the mold in the proper thicknesses. Each sheet was about 1/8 inch thick. Up to four layers were needed in spots to be forced into the irregularities in the mold shape. These irregularities included outside body details and interior bosses for attachment screws. The consistency of the material was about like a child's modeling clay, although it held together in sheets because of the chopped fiber. After the compression and heating, the halves of the truck hoods were a rigid but somewhat flexible solid material.

The new method of joining material as applied to these two sides of the truck hood was adhesive bonding along a lap joint along the centerline of the hood. The lap joint is typically at least an inch wide. The adhesive is supposed to spread throughout the joint area when the two parts are brought together, and then is supposed to cure, while the parts are held together. A schematic representation of a lap joint is shown in Figure 9.3. The quality assurance method available to the customer (truck manufacturing company) was a sheet metal shim to probe the lap joint from the visible side to determine whether adhesive was present. The customer had bought off on this approach. The real situation was more complex, as the adhesive could be present, absent, well-adhered, or poorly adhered.

The two sides of the hoods were molded by a first-tier supplier. This supplier also did the adhesive bonding of the two sides into a complete hood. Then the supplier shipped the hoods to the purchaser's truck assembly

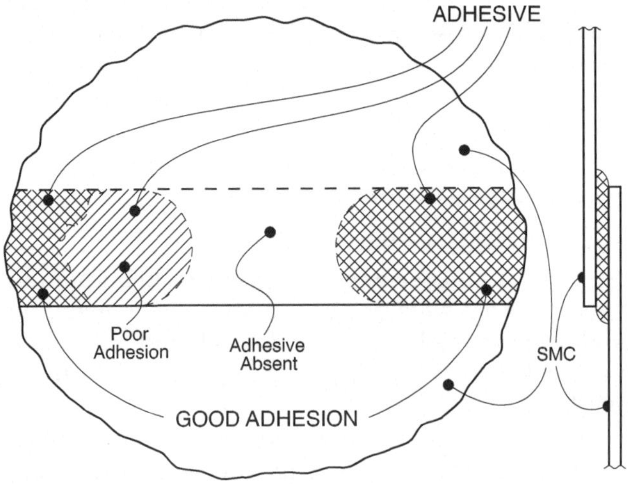

FIGURE 9.3
Schematic representation of an adhesive lap joint. As noted, the adhesive may be present, absent, well adhered, or poorly adhered. A sufficient amount of adequately well-adhered adhesive was desired per unit length of the lap joint.

plant for final assembly into vehicles. Failures of these adhesive bonds were finally reported to the truck manufacturer. Only then was the NDT development group called in.

Failures of the adhesive bond can be caused by (1) unclean surfaces, (2) lack of adhesive, (3) precure of the adhesive if the parts are not put together soon enough, and (4) spring-back of the parts if they are not clamped into position during the curing process.

The problem was compounded by all of these causes, not just one. Contamination could never be ruled out because of the handling routine. Adhesive was applied by hand with things like caulking guns so that areas could be missed in a rush situation. Workers could take a cigarette break between the application of the adhesive and the joining of the parts, letting the adhesive begin to cure. Because the parts were not clamped but simply set aside, gravity and mismatch could cause parting of the adhesive line during curing at room temperature. And, compounding the problem still further, a relatively rapidly polymerizing adhesive was used so that the parts would not have much time to sag apart before curing. This attempt to circumvent the spring-back problem (without the use of clamping jigs) exacerbated the precure problem if there were assembly delays. This analysis is sort of a FMEA after the fact. The root cause of the problem was failure to follow W. E. Deming's Point 4: end the practice of awarding business on price tag alone. The hood supplier had been the low bidder on the truck hood job.

The problem showed in the field where fleets of new trucks were falling apart. Failure rates up to 40% were experienced. Because these heavy trucks were supposed to be durable for industrial jobs, the truck manufacturer's reputation was on the line. To complicate the situation, the first-tier supplier was secretly repairing adhesive bonds in the field without informing the warranty division of the truck manufacturer. However, the supplier was eventually caught. The truck manufacturer calculated the actual loss at \$250,000 per year plus a large multiple for damage to reputation. This dollar figure provided an opportunity to use integrated costs for a year in the DIC calculation below. Specifically, $\Sigma(k_2 \times p) = \$250{,}000$ over one year.

The most obvious solution—to change processes or to change suppliers—was complicated by contractual obligations and the time to renegotiate and plan, probably 2 years. The situation was so bleak that the truck company management had issued an edict (manufacturing feasibility rejection) declaring the use of adhesively bonded SMC parts to be infeasible in manufactured products. The next step would have been an order to stop production, bringing heavy-truck production to a screeching halt. The threat of this action was real and its implementation was rapidly approaching. A return to steel bodies would have been next to catastrophic.

At that point in time, an NDT inspection method was recognized to be necessary, but none was available. The truck company wanted to be able to inspect bonded truck bodies as they arrived at the assembly plant, and to retrofit such inspection into the first-tier supplier's plant. The truck manufacturing company wanted a field-portable method for obvious reasons.

As stated above, the only test method available to the truck company was a gross test for the absence of adhesive. A feeler gage shim was used as a probe between the two layers of SMC to detect whether adhesive was missing. This test proved ineffectual because many truck hoods were observed with the edges of adhesive joints "buttered over" with extra adhesive, which prevented the entry of the shim. Sawing up these hoods revealed that the adhesive was missing from within the joints. Besides, the shim method did not address the question of weak bonds containing an inadequate amount of adhesive (i.e., poor adhesion).

The plastics design group of the truck company assembled a task force and looked up as many NDT methods and instruments as they could find, but found no definitive answers in off-the-shelf products. They came to me to evaluate these leads or invent a new method. At the time I was head of the NDT research, development, and applications group.

One senior engineer was assigned to the job and he singled out one suggested ultrasonic instrument as having some potential. This was the Sondicator Mk II manufactured at the time by Automation Industries (now redesigned by Zetek). The Sondicator used Lamb waves at approximately 25 kHz propagating between two closely spaced probe tips. The instrument is hand portable, about 20 cm × 20 cm × 30 cm, with a shoulder strap. The configuration of this probe on a lap joint is shown in Figure 9.4. Actually, the wave motion involved both propagating waves and evanescent waves

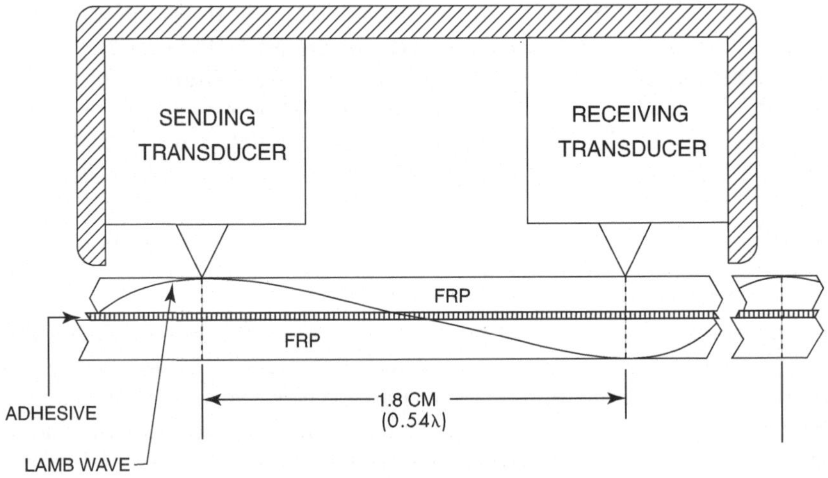

FIGURE 9.4
Schematic representation of the ultrasonic Lamb wave probe on a lap joint. The wave motion is partly a traveling wave and partly an evanescent wave around the input tip. The phase and amplitude of the received signal is compared with the input by the attached instrument. (From Chapman, G. B. II, Papadakis, E. P., and Meyer, F. J. (1984). "A Nondestructive Testing Procedure for Adhesive Bonds in FRP Assemblies," *Body Engineering Journal*, Fall, 11–22. With permission. Copyright 1984 © Open Systems Publishing.)

analogous to resonance near the tips. The received signal was compared in both amplitude and phase to the input signal by means of built-in circuitry, and poor bonds were signaled by a red light and an audible tone burst. The Sondicator required calibration against acceptable reference standards of adhesively bonded material.

The Sondicator was immediately found to be capable of detecting the difference between well-adhered adhesive in the lap joints and the lack of adhesive over moderate areas including "buttered-over" vacant regions. However, further work was required to detect present but not-adhered adhesive, and adhesive with weak bonds.

The engineer made a breakthrough on this question by making one important discovery. The Sondicator would reject almost all industrially made bonds if it was calibrated against perfectly made bonds in the laboratory. In reality, many of the industrially made bonds were strong enough to survive in the field. The test in this stage of development would have rejected all of production. The engineer's conclusion was that the "perfect" laboratory calibration standard was worthless. It followed that he had to create a calibration standard containing the requisite degree of imperfection to just barely accept the acceptable bonds and reject the bonds that were actually made but were unacceptably weak.

The engineer solved the problem of creating sufficiently imperfect reference standards by applying statistics to a large family of bond samples made in the supplier's factory by hourly personnel under production conditions. These samples were tested and rank ordered with the Sondicator modified to give quantitative read-outs, not just the red light and tone burst "no-go" alarm of its regular operation. Physical tensile pull-tests then determined the Sondicator level corresponding to the rejectable strength level. The reference standard was born as the type of sample just good enough to exceed the minimum specifications of the pull-test. With the reference standard, the no-go test could be used.

At this point it was possible to compute all the costs for the DIC calculation. The testing cost for a year at the truck plant incoming area was \$3,000 for the instrument plus \$25,000 in variable costs, principally for labor, adding up to $\Sigma(k_1)$ = \$28,000 over a year. The not-excessively-good standards had already been made in the laboratory. The detrimental cost for the year was set at \$250,000 = $\Sigma(k_2 \times p)$ found earlier. The value of the proportion p nonconforming was actually 0.40 like the experience in the field, but is not used directly. The resultant value of the DIC using Equation 7.1 is

$$\begin{aligned} \text{DIC} &= \sum (k_2 \times p) / \sum (k_1) \\ &= (\$250{,}000)/(\$28{,}000) \end{aligned}$$

so that

$$\text{DIC} = 8.93 \tag{9.2}$$

The inspection was instituted.

The development engineer then taught the method at the plant where the trucks were assembled. The technology transfer was performed seamlessly. The truck company also instructed the first-tier supplier on the use of the method so that high quality could be ensured at the supplier and so that nonconforming product would not be shipped to the assembly plant.

The quality management office of the truck manufacturer accepted the method after the development engineer wrote it up in the standard format. The method then served to define a specification for an adequate adhesive lap joint on a per-unit-length basis. No such specification had existed in the industry previously. The engineer's new specification (Ford Motor Co., 1980) is now accepted as an exact parallel to the spot-weld specification for steel.

The edict declaring adhesively bonded SMC to be infeasible in a manufacturing context was rescinded just weeks before the order to stop truck production was to have been issued. One can imagine the magnitude of disruption that would have occurred if the company had been forced to revert to steel truck bodies. It would have affected the plastics industry, the company's stamping plants, steel sheet orders, fuel economy, corrosion lifetimes of bodies, and all the future designs for a variety of SMC parts for additional trucks and cars. As the feasibility of adhesive bonding of SMC was reestablished, the use of SMC was extended to other parts and other car lines, thus improving corporate average fuel economy (CAFÉ) mileage and durability with respect to rust. The rescuing of SMC and the elimination of all the above problems is directly attributable to NDT applied with imagination and the requisite degree of smarts.

The first-tier SMC supplier reduced its failure rate from 40% to around 5% simply because it became cognizant that it could be monitored by NDT. Other parts went into production in later years because their bonding quality could be assured. NDT paid for itself many times over. Continued calculations showed that the DIC remained higher than 1.0. The inspection method remained a requirement to ensure that the specification for lap joints was being met.

The method developed by the development engineer is written up in his articles (Chapman, 1981, 1982a, 1982b, 1983, 1990, 1991; Chapman and Adler, 1988; Chapman et al., 1984; Maeva et al., 2004; Meyer and Chapman, 1980; Papadakis and Chapman, 1991; the financial analysis is given in Papadakis, 1985 and is used here).

Choosing bidders on price alone is bad, but doing so without methods to test their wares for latent defects is even worse.

The Deming inspection criterion was applied successfully to prove that testing should be done on adhesive lap joints in SMC parts.

9.2.3 A Safety-Related Part: Front Wheel Spindle Support

It was stated previously that 100% inspection of all safety-related parts is essential under Deming's philosophy and under the protocols advocated by NDT experts. For safety-related parts, k_2 in the DIC approaches infinity, so the DIC is always much larger than 1.0, requiring testing. The interesting fact about the safety-related part to be described here is that a scenario could be worked out in which the use of batch sampling could have been adequate. The decision to use 100% testing by ultrasonic velocity is convoluted and should be studied and understood.

The part is a front wheel spindle support for a rear-wheel drive vehicle. Many other parts such as brake caliper brackets, for instance, are also treated this way. There is a right and a left front wheel spindle support. This part holds the wheel spindle, which is press-fit into the spindle support. The spindle is actually a stubby axle for the individual front wheel. The spindle support is attached to the McPherson strut, which contains the spring and the shock absorber. The attachment mechanism is a pivot that permits steering. The spindle support also has an arm to which the steering push-rod is attached. The braking mechanism is also attached to the spindle support. As one can well imagine, a failure of a front wheel spindle support could have dangerous consequences.

The front wheel spindle supports in this case are made of nodular iron cast by the batch process from large inoculated ladles. High nodularity is required. As explained in Section 9.2.1, all the parts from a ladle (batch) are good if the last one cast is good. To be able to ensure that the batch is all good, the batch must be kept together until the last part made can be tested. Because many batches are made during each shift in a casting plant, and because it is much easier to intermingle batches and not keep them separate, it was decided to test every part instead of trying to maintain batch traceability to do batch quality assurance. The net cost of doing 100% inspection was calculated to be cheaper than the process of keeping batches separate. The inspection is performed by ultrasonic velocity as described in Chapter 8. Thus, 100% inspection is mandated in the car company's own casting plants and in the foundries of its suppliers. Thus the safety requirement of having high nodularity in all critical parts is met by 100% inspection. The use of 100% inspection in this case is consistent with Sections 7 and 8 of ISO-9000–2000, as explained in Chapter 5.

9.2.4 Several Identical Parts in One Subassembly: Connecting Rods

Nodular iron connecting rods were being planned for substitution in place of forged steel connecting rods in six-cylinder automobile engines. The automated casting machine gave adequate nodularity for strength, but the castings displayed voids on occasion. The failure rate p was found to be 0.01 or greater from experience in early production. Six good connecting rods were needed for each engine. The NDT method of choice was x-ray fluoroscopy read in real time by operators. The connecting rods were carried on a moving belt. Operators at the NDT vendor company were spelled, 15 minutes on and 15 minutes off, to avoid visual fatigue. They were doing visual testing (VT), although the technology was x-ray. The DIC calculation was used to justify the NDT. To replace an assembly (engine) upon failure, the cost k_2

was approximately \$1000. Because six parts had to survive simultaneously in each engine, the gross value of p to account for six good parts is

$$p \approx [1 - (0.99)^6] \approx 0.06 \tag{9.3}$$

The value of k_1 to test six parts by the VT/x-ray method was quoted by the NDT vendor company at \$2.20. The value of DIC is

$$\begin{aligned} \text{DIC} &= k_2 \times p/k_1 \\ &= (\$1000 \times 0.06)/(\$2.20) \end{aligned}$$

so that

$$\text{DIC} \approx 27 \tag{9.4}$$

If customer loyalty were to add another \$1000 to k_2, then DIC would be 54. The automobile company decided to institute the NDT.

After 2 years on a learning curve, k_1 was reduced to \$0.90 per six parts. At the same time p per part stabilized in a range between 0.00375 and 0.0050 because of process improvements (continuous improvement). With k_2 at \$1000 without customer loyalty contributions, the value of DIC is still 4.1 to 5.5, indicating the continued need to inspect 100%. Inspection was continued.

9.2.5 Intermediate Inspection of a Machined Part: Engine Block

The engine division of an automotive company received raw castings of V8 engine blocks from the casting division of the company. After machining, nonconforming blocks were returned as scrap to the casting division, which supplied fresh raw blocks, one-for-one, at no charge. Thus, from the engine division point of view, the only loss was its machining costs.

One flaw that showed up to make a block nonconforming after machining was porosity in the regions that would become cylinder walls. The machining opened up the porosity into holes. These holes were discovered 100% at the end of the cylinder bore honing operation by pressure decay testing and by visual testing by operators using flashlights and dental mirrors to look into the cylinder bores. No further detrimental costs were associated with these flaws. The pressure testing and visual testing were to continue no matter what the results of the proposed NDT.

NDT was called in to determine whether any machining costs could be saved by installing automatic electronic testing earlier in the manufacturing line. There were three steps: rough machining, fine machining, and honing. The raw castings were too irregular to interrogate electronically. The NDT probes could be used after rough machining, saving the cost of fine machining and honing. The engine plant controller provided the cost as \$8.97 for the fine machining and honing of the cylinders of one V8 engine block. This is the value for k_2 in this case.

Two estimates for NDT equipment were received—$120,000 and $170,000. As the production of the type of engine was to be extended over an uncertain number of years, the calculations of savings by means of this type of equipment were desired over 1, 3, 6, and 10 years. For a large investment in equipment, the formula for k_1 is approximated by

$$k_1 \approx [(I - R + C)/(L \times V)] \tag{9.5}$$

where

I = investment,
R = residual value of test equipment,
C = operating cost totaled over life cycle L,
L = life cycle of the test,
V = average production volume in one year.

The proportion of defective parts, p, for the DIC calculation must be the part of the total nonconforming material caught by the test. It was known from historical data that the scrap rate due to the kind of flaws to be detected was between 8,500 and 15,000 cylinder blocks per year out of a production volume of 300,000. The instruments quoted could probably detect 60 to 80% of flaws of that size. Hence the effective value of p became the probability of detection times the scrap rate divided by the production volume yielding values of p between 0.017 and 0.040. Using reasonable values of all the variables, a family of 72 calculations was carried out to yield the DIC over 1, 3, 6, and 10 years. Values of the DIC varied from 0.5 to 7.5.

One would have normally concluded that for some sets of probable input parameters, the DIC would have indicated the propriety of investing in the equipment and instituting the test, but this case history was more complicated. The company was beset by the nationwide double-digit inflation of the time, which cut the sales of automobiles and lowered the profitability of investments. A hurdle rate of 52% had been established by the treasurer for investments in equipment. It was necessary to carry out a calculation of time-adjusted rate of return (TARR) for the investments of $120,000 and $170,000 to find out if the return would be higher than the hurdle rate and if the payback period would be short enough. This TARR calculation will be reported in Section 9.3 on examples of the internal rate of return (IRR) and TARR methods.

9.3 Examples of TARR and IRR Methods

These examples were first presented in various articles and short courses by the author and some colleagues. Prior to giving real-world examples, a didactic case will be presented with simple numbers to show the principle of the TARR and IRR finances.

9.3.1 Didactic Example: Hypothetical Data

These financial methods involving investments calculate the rate of return the test method allows on borrowed money to purchase the equipment. Without the test, the organization would accrue some detrimental cost. This cost is saved by installing the inspection method and becomes the revenue for the calculation. Costs and revenue are written down year by year. The investment is a negative quantity at year zero. Depreciation is a negative quantity each year to calculate the profit for the year, but is added back in to get the cash flow for that year. Operating costs and maintenance fees are on the negative side. The cumulative cash flow is added up each year and becomes the data for the IRR and TARR calculations. The company or factory controller will have software for these calculations. The IRR calculation is available in many mathematical, engineering, and accounting software packages.

A simple hypothetical case is tabulated in Table 9.1. The initial investment is taken as $50,000. This equipment is used 10 years with operating costs of $3,000 per year and maintenance of $1,000 in alternate years. Linear depreciation is $5,000 per year. The savings arise from warranty costs of $20,000 per year. The profit in a year with no maintenance is $12,000 from savings when operating costs and depreciation are subtracted. (In the alternate years it is $11,000.) With $12,000 taxed at 50%, the remainder is $6,000. Adding back the depreciation leaves cash flow of $11,000. For alternate years the cash flow is $500 less, but is rounded up to $11,000 for convenience. In the

TABLE 9.1

Hypothetical Data Illustrating IRR and TARR ($ thousands)

Year	0	1	2	3	4	5	6	7	8	9	10
Investment	(50)										
Depreciation		(5)	(5)	(5)	(5)	(5)	(5)	(5)	(5)	(5)	(5)
Residual value											10
Operating cost		(3)	(3)	(3)	(3)	(3)	(3)	(3)	(3)	(3)	(3)
Warranty cost		20	20	20	20	20	20	20	20	20	20
Maintenance			(1)		(1)		(1)		(1)		(1)
Pretax profit		12	11	12	11	12	11	12	11	12	21
Taxed 50%		6	5.5	6	5.5	6	5.5	6	5.5	6	10.5
Cash flow,[a] rounded	(50)	11	11	11	11	11	11	11	11	11	16
Cumulative cash flow	(50)	(39)	(28)	(17)	(6)	5	16	27	38	49	65

[a] Depreciation added back in to the after-tax profit indicate the cash available to the firm.

Source: Papadakis, E. P., Stephan, C. H., McGinty, M. T., and Wall, W. B. (1988). "Inspection Decision Theory: Deming Inspection Criterion and Time-Adjusted Rate-of-Return Compared," *Engineering Costs and Production Economics*, 13, 111–124. With permission from Elsevier.

last year on the 365th day the equipment is sold for its residual value, which turns out to be $10,000 on the open market, and the cash flow with that extra income taxed is $16,000. The last row in Table 9.1 is the cumulative cash flow to input into the IRR software.

The result for this hypothetical set of data is a TARR of 18% and a payback period of 4.5 years.

The cash flow being compared was shown in Figure 7.2 on the theory of the IRR and TARR methods. Please refer back to that illustration. One is contrasting the nontesting case (the present situation), which has high warranty repair detrimental costs, vs. the inspection case with its other set of costs. Any scenario for considering the institution of a test can be laid out this way.

9.3.2 Intermediate Inspection of a Machined Part

The machined part is the same engine block treated in Section 9.2.5. The family of 72 calculations was performed for the TARR as well as for the DIC as given above. The results for the four process lifetimes are shown in Table 9.2, where

TABLE 9.2

DIC, TARR, and Payback Periods for All the Sets of Calculations on Engine Blocks

L = 1 Year			3 Years			6 Years			10 Years		
DIC	TARR	Payback	DIC	TARR	Payback	DIC	TARR	Payback	DIC	TARR	Payback
									0.793	2.63	—
									1.005	−0.81	—
			0.623	−3.22	—	0.795	−2.31	—	1.047	−0.4	—
			0.934	−0.74	—	0.993	−0.65	—	1.053	−0.47	—
0.522	−2.62	—	1.027	−0.16	—	1.013	−0.47	—	1.058	−0.41	—
0.782	−1.28	—	1.152	0.79	3.96	1.033	−0.30	—	1.349	1.86	10.27
1.017	−0.10	—	1.183	1.03	3.94	1.192	1.02	6.86	1.587	3.64	9.13
1.043	0.04	2.00	1.245	1.49	3.92	1.589	4.16	6.41	2.116	7.21	7.13
1.304	1.36	1.97	1.557	3.79	3.79	1.986	7.15	5.91	2.645	10.44	5.91
1.565	2.69	1.94	1.868	6.09	3.66	2.384	10.06	5.06	3.174	13.44	5.06
1.825	4.01	1.91	2.179	8.33	3.53	2.781	12.80	4.46	3.703	16.20	4.46
2.086	5.32	1.88	2.491	10.55	3.41	3.178	15.44	3.98	4.232	18.79	3.98
2.114	5.46	1.88	2.524	10.78	3.39	3.221	15.71	3.94	4.289	19.06	3.94
2.467	7.26	1.84	2.945	13.75	3.22	3.758	19.13	3.46	5.004	22.35	3.46
2 819	9.03	1.80	3.366	16.64	3.04	4.295	22.36	3.08	5.718	25.42	3.08
2.758	8.72	1.81	3.292	16.14	3.07	4.201	21.81	3.14	5.594	24.89	3.14
3.217	11.01	1.76	3.841	19.82	2.76	4.901	25.81	2.76	6.526	28.66	2.76
3.677	13.32	1.70	4.390	23.42	2.45	5.602	29.62	2.45	7.459	32.22	2.45

Source: Papadakis, E. P., Stephan, C. H., McGinty, M. T., and Wall, W. B. (1988). "Inspection Decision Theory: Deming Inspection Criterion and Time-Adjusted Rate-of-Return Compared," *Engineering Costs and Production Economics*, 13, 111–124. With permission from Elsevier.

the production volume is 300,000 and the investment is $170,000 for the equipment judged to have the higher probability of success. One can see that the value of the TARR never exceeds 33%. In the economics of the time (1987) with the hurdle rate set at 52% by company management, the NDT test was not justifiable despite the high values of the DIC for some combinations of data. It was decided to refrain from initiating the inspection on all the blocks to save the machining cost on blocks that would ultimately be scrapped.

The calculated results in Table 9.2 are instructive in showing the relationship between the DIC and the TARR. The values of DIC are plotted against the values of TARR in Figure 9.5 for each of the four product lifetimes. The curves are monotonic increasing as expected. Also, as expected, the curves all pass through the point (TARR = 0.0, DIC = 1.0). This means that at DIC = 1.0, which is the breakeven point, the interest rate at which equipment would have to be purchased would be zero. Breaking even is earning no interest, so the situation means that the two financial theories agree at that single point. This result is theoretically satisfying and useful in practice in convincing financial people of the mutual validity of the theories.

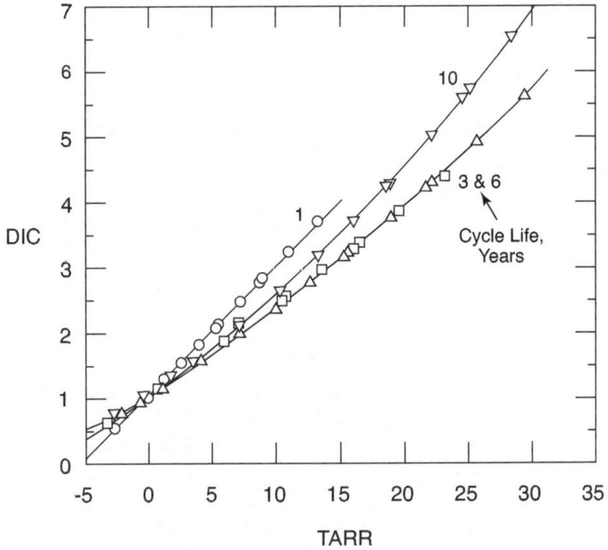

FIGURE 9.5
DICs are plotted against TARR for each of the four product lifetimes. (From Papadakis, E. P., Stephan, C. H., McGinty, M. T., and Wall, W.B. (1988). "Inspection Decision Theory: Deming Inspection Criterion and Time-Adjusted Rate-of-Return Compared," *Engineering Costs and Production Economics*, 13, 111–124. With permission.)

9.3.3 Aircraft Engine Discs

Aircraft jet engine turbine discs have experienced failures arising from cracks. One example was the failure of a disc in the rear engine of a Douglas DC-10 over Iowa in 1989. The broken disc destroyed the hydraulic lines and disabled the flight control surfaces of the aircraft. A crash landing resulted in loss of life of many on board.

When a disc fails at an engine speed of around 20,000 rpm, it generally breaks into three parts flying outward at high speed. The pieces pierce the shroud of the engine, the nacelle, and whatever part of the airplane they may be traveling toward. One or more of the parts may travel away from the plane in the air. The parts that hit the plane do major damage and may destroy the airplane, as in the example cited.

Turbine discs are made of titanium and have a theoretical lifetime calculated by crack growth and fracture mechanics equations. They are retired and replaced after this lifetime, which depends upon the number of times the engine is brought up to maximum rpm. This is a fatigue failure mechanism depending upon repetitive stresses. The Air Force has kept thousands of discs that have outlived their design lifetimes in hopes of developing an NDT that would permit life extension of the discs.

The fatigue cracks due to stress cycling are one of the latent defects spoken of by Dr. Deming. Generally the cracks arise from internal defects not visible on the surface. These are usually inclusions in the original metal that become the source of cracks during the long period of fatigue. The cracks start out small and interior; they must grow larger and reach the turbine disc surface before becoming dangerous.

The titanium originally comes from castings that are forged into ingots by a metallurgical supplier. The ingots are sliced into relatively thin sections and then forged into discs of near net shape by the engine manufacturer. There is machining work done to finish a disc. The perimeter of the disc has "Christmas tree" dovetail notches to hold the turbine blades. Discs are considered *related*, for recall purposes, if they are related to each other by the prior metallurgy and manufacturing occurrences. For instance, all discs manufactured from one ingot would be considered related.

The inclusions are different from the parent metal, so the metal receives greater stresses around an inclusion in the forging process. It is reasonable to expect inclusions to become sources of cracks. It is the practice among jet engine manufacturers to test the discs with the best state-of-the-industry NDT equipment during manufacture to discover and destroy the discs that have inclusions. For internal flaws such as inclusions, ultrasound is used. A system using immersion and computer-controlled aiming of the ultrasonic probe scans the entire interior of the part. Echoes from the interior indicate an inclusion or a void.

The turbine discs are shot-peened to put a compressive stress into the surface layer. This compressive layer retards the growth of any cracks that may start at the inclusions because the cracks can propagate only when the metal at the crack tip is under tension. The acceleration and centrifugal force as the engine gets up to speed tends to produce tensile forces that permit crack growth. The tensile forces are counteracted by the compression left over from the shot-peening. The shot-peening is sufficient to retard the crack growth to the surface for a period of 6 to 8 years. Thus, failures will not be seen until the sixth year after manufacture in these discs and engines with typical operation.

The metallurgical suppliers have been carrying on continuous improvement for many years to attempt to lower the number of inclusions in an ingot. Based on the small number of turbine discs discovered to have inclusions in certain years, the statistics-based quality professionals at one engine company began to advocate termination of the 100% ultrasonic inspection that discovered the inclusions. The NDT engineer at the company opposed the proposed termination. The company was a member of the Center for Nondestructive Evaluation at Iowa State University at the time, and the NDT engineer was the delegate from the engine company to the center. In this capacity, the NDT engineer (Bray, 1990) brought the turbine testing question to the center where I was an associate director. The NDT engineer wanted financial proof that the testing should continue. The turbine testing question turned out to be a perfect test case for the internal rate of return calculation. The financial data were collected as follows.

Production and discontinuity data were available for the years 1983 through 1988 inclusive. The data (Howell, 1990) are given in Table 9.3. These data were published earlier (Papadakis, 1995). The discontinuities detected by ultrasound were confirmed optically by cutting the detected discs and examining the cut surfaces. Each ultrasound indication resulted in the discovery of an inclusion. One inclusion too near the surface for a definitive ultrasound indication was discovered by visual inspection after final machining had opened it to the surface. This discontinuity is not included in the financial calculation to prove the efficacy of the ultrasound inspection.

TABLE 9.3

Production and Discontinuity Data on Aircraft Jet Engine Turbine Discs

Year	Production	Discontinuities Found by Ultrasound
1983	2,516	12
1984	4,541	9
1985	6,523	4
1986	6,222	3
1987	2,663	1
1988	3,500	5
Total	25,965	34

Source: Papadakis, E. P. (1995). "Cost of Quality," *Reliability Magazine*, January/February, 8–16. With permission from Industrial Communications, Inc.

The investment in the computer-controlled ultrasonic immersion flaw-detection system had been made at the end of 1982, which becomes year zero for the IRR calculation. The cash flow at year zero is the cost of the system taken as a negative quantity, ($400,000). The turbine discs are all of the same type but not all of the same size. The variable cost to test a disc ranged form $11 to $22 with the weighted average being $17. Depreciation was not included because it was expected at the outset that the equipment would be adapted to new models by reprogramming the steering computer. The cost per year for testing for 1983 through 1988 inclusive is $17 times the annual production from Table 9.3. These costs are taken as negative in the cash flow. If the testing had not been carried out, damage to aircraft would have started in the sixth year after the first year's production—1989. According to the theory of fatigue, the failures from the first year's production would have been spread out over time 6, 7, and 8 years later (1989, 1990, and 1991). Similarly, the failures from the second year's production would have been spread out over 1990, 1991, 1992, and so on.

The savings due to testing arise from the avoidance of the damage to aircraft that would have been caused by these nontesting failures. The historical data available to the engine company were that the cost of the destruction of an engine and a nacelle would have been $500,000 in liability, while the cost of the destruction of a plane would have been $7,000,000. The data were for a plane on the ground with no injuries. A crash from 35,000 feet with two or more fatalities would have been much more costly. The NDT engineer was satisfied to use the $7 million figure in the IRR calculations.

It was necessary to construct a hypothetical distribution of failure dates for flawed parts from each production year to account for the delays of 6, 7, or 8 years before fatigue failures could occur. Then the failures in particular years could be postulated and the cash flows due to warranty savings could be calculated. The failures vs. years are shown in Table 9.4. The real-world occurrences would be only marginally better or worse.

The costs were calculated from year zero onward for two scenarios, the destruction of airplanes and the destruction of engines and nacelles. The costs are listed in Table 9.5 as worst case and best case, respectively. The costs were used as input data into IRR software. The results for the IRR are shown as the last entries in the columns in Table 9.5. The IRR for the case of destroying airplanes is 105.65%, and the IRR for the case of destroying engines and nacelles is 46.62%. As the hurdle rate of the engine company was 12% at the time of the calculation, the investment was eminently justified.

However, the above calculation should be considered one stage beyond the worst case for safety reasons. In the real world, if failures began to occur, the engine company would mount recall campaigns on the related discs. The company data showed that the cost of such a campaign was $8,000,000. A campaign would have replaced many of the flawed discs manufactured the same year as the flawed disc that failed in service. As the flawed discs were actually discovered and destroyed in the year of manufacture, the NDT engineer was not given any data on which ones were related. Hence, the

TABLE 9.4

Hypothetical Failures vs. Years for the Flawed Parts in the Production of Turbine Discs

	Production Year	**1983**	**1984**	**1985**	**1986**	**1987**	**1988**	**Total**
	Detected Flawed	**12**	**9**	**4**	**3**	**1**	**5**	**34**
Year	**Year of Failure**				**Number Failing**			
0	1982							
1	1983							
2	1984							
3	1985							
4	1986							
5	1987							
6	1988							
7	1989	4						
8	1990	4	3					
9	1991	4	3	1				
10	1992		3	2	1			
11	1993			1	1	0		
12	1994				1	1	1	
13	1995					0	2	
14	1996						2	
	Total	12	9	4	3	1	5	34

Source: Papadakis, E. P. (1995). "Cost of Quality," *Reliability Magazine*, January/February, 8–16. With permission from Industrial Communications, Inc.

TABLE 9.5

Cash Flows vs. Years for the Flawed Parts in the Production of Turbine Discs

Year	**Worst Case Plane Destroyed**	**Best Case Nacelle Destroyed**
0	(400,000)	(400,000)
1	(42,772)	(42,772)
2	(77,197)	(77,197)
3	(110,891)	(110,891)
4	(105,774)	(105,774)
5	(45,271)	(45,271)
6	(59,500)	(59,500)
7	28,000,000	2,000,000
8	49,000,000	3,500,000
9	56,000,000	4,000,000
10	42,000,000	3,000,000
11	14,000,000	1,000,000
12	21,000,000	1,500,000
13	14,000,000	1,000,000
14	14,000,000	1,000,000
	IRR	
Result	105.65%	46.62%

Source: Papadakis, E. P. (1995). "Cost of Quality," *Reliability Magazine*, January/February, 8–16. With permission from Industrial Communications, Inc.
Note: Parentheses indicate negative numbers.

TABLE 9.6

Posited Failures Assignments Accounting for Campaigns

	Production Year	**1983**	**1984**	**1985**	**1986**	**1987**	**1988**	**Total**
	Detected Flawed (From Sets)	**12(2)**	**9(2)**	**4(1)**	**3(1)**	**1(1)**	**5(1)**	**34(8)**
Year	**Projected Year of Failure**	**Number Failing**						
0	1982							
1	1983							
2	1984							
3	1985							
4	1986							
5	1987							
6	1988							
7	1989	1						1
8	1990	1	1					2
9	1991		1	1				2
10	1992				1			1
11	1993					1		1
12	1994						1	1
	Adjusted Total	2	2	1	1	1	1	8

Source: Papadakis, E.P. (1997b). "A Cost of Quality: Three Financial Methods for Making Inspection Decisions." Materials Evaluation 55(12), 1336–1345. With permission.

recalls included in the next calculation were hypothetical. To arrive at a failure distribution, many of the failures posited in Table 9.4 were set equal to zero to indicate that the offending discs had been eliminated in the recalls. The resulting posited failure occurrences per year are given in Table 9.6 (Papadakis, 1997b).

Another cost table like Table 9.5 was constructed. The testing costs in the years of manufacture were the same, of course. The warranty costs were different. Every time there was a failure in Table 9.6, the cost was either $7 million for a plane in the worst case or $500,000 for a nacelle in the Best Case. However, to each of these was added $8 million for the recall of related parts to lower the number of failures from the list in Table 9.4 to the list in Table 9.6. The resulting costs are shown in Table 9.7.

The costs in Table 9.7 were used in the IRR software to produce the IRR results in the last line of Table 9.7. The high IRR shows that the inspection should be continued. The engine company did, indeed, continue the inspection on 100% of production of jet engine discs.

The calculation shows that inspection can be profitable. The conventional wisdom is that inspection is an expense to be eliminated by improving manufacturing techniques. Of particular significance to the question of profitability is the fact that the IRR is large and positive in this case, even when the process capability is high. In certain cases like this where continuous improvement has been carried out diligently, the need for NDT applied to 100% of production still remains. The few remaining parts with

TABLE 9.7

Cash Flows vs. Years with Campaigns Included

Year	Worst Case Plane Destroyed	Best Case Nacelle Destroyed
0	(400,000)	(400,000)
1	(42,772)	(42,772)
2	(77,197)	(77,197)
3	(110,891)	(110,891)
4	(105,774)	(105,774)
5	(45,271)	(45,271)
6	(59,500)	(59,500)
7	15,000,000	8,500,000
8	30,000,000	17,000,000
9	30,000,000	17,000,000
10	15,000,000	8,500,000
11	15,000,000	8,500,000
12	15,000,000	8,500,000
	IRR	
Result	90.53%	77.06%

Source: Papadakis, E.P. (1997b). "A Cost of Quality: Three Financial Methods for Making Inspection Decisions." Materials Evaluation 55(12), 1336–1345. With permission.

Note: Parentheses indicate negative numbers.

nonconformities must be detected and eliminated by inspection in order to preclude potential catastrophes and to eliminate the concomitant large adverse costs. Cost avoidance provides the computed profit. The investment in high-tech inspection technology equipment is needed to detect the latent defects that yield their adverse costs over time.

The same sets of data will be used to illustrate the application of the productivity method that follows the trail of productivity, profitability, and revenue.

9.4 Examples of the Productivity, Profitability, and Revenue Method

9.4.1 New Metal for Automotive Connecting Rods

Sintered and coined powder metal (P/M) was to be introduced as a new material for connecting rods in an I4 automobile engine (in-line four-cylinder engine). The author was called in as the NDT expert during the concurrent engineering phase 2 years before production was to begin. In consultations, the chief metallurgical engineer on the connecting rod development project, Stan Mocarski, presented the following projections (Mocarski, 1983) on production and costs for my use in NDT cost–benefit calculations. As will be seen below, the benefits far outweighed the costs. My group initiated the development work on an NDT inspection method immediately, working closely with the others in the concurrent engineering effort. The development was successful. Its final use after a circuitous route of implementation was reported earlier in Chapter 4, Section 4.2.9. The projected costs and production figures are used in the productivity, profitability, and revenue calculations that follow.

The planned production volume was one million engines in the first year. This translates to 4,000,000 connecting rods in the year. Experience from the development process indicated that the proportion defective could be expected to be 1 in 10,000, or $p = 0.0001$. Over one year this would represent 400 nonconforming rods. It is assumed that the production of nonconforming parts is random so that no more than one nonconforming connecting rod would be in any one engine. A nonconforming connecting rod could fail and destroy an engine. The warranty replacement price for a failed engine would be \$15,000. Other production costs were given as follows: The production price per rod is \$5.00, while the price to produce rods at overtime rates to replace nonconforming rods when detected is \$10.00. The transfer price of a completed rod from the rod machining area in the factory to the engine assembly area is \$10.00. (The above four cost figures are estimates before Job 1 and the initiation of any learning curves.) The cost to do 100% inspection by eddy currents on the connecting rods to detect the types of failures experienced and predicted by FMEAs is \$200,000 per year. Further, it is assumed that only half of the nonconforming rods will fail under the 12/12 warranty offered (12 months or 12,000 miles of operation, whichever comes first).

To use the productivity method, Equations 7.3 through 7.5 must be implemented. In this example, one must do the implementation in three scenarios and compare the results. The scenarios are as follows:

- Baseline of perfect production with no testing and no nonconforming parts produced. There is no warranty cost and no overtime for replacement parts.
- Production of nonconforming parts but no inspection. Warranty enters but there is no overtime.
- Production of nonconforming parts, but inspection eliminates all of them. Warranty is zero. Replacement parts require overtime, and inspection adds to production cost.

9.4.1.1 *The Baseline Calculation*

The first quantity calculated is the value of A, which in this case is the number produced times the transfer price, or $A = N \times T = 4{,}000{,}000 \times \$10.00 = \$40{,}000{,}000$. There are no contributions to the disvalue (value-added detractor), so $B = \$0$.

The value of C is the number produced times the production cost, or C = 4,000,000 × \$5.00 = \$20,000,000. The resulting productivity is

$$P = (A - B)/C$$
$$= (\$40{,}000{,}000 - \$0)/(\$20{,}000{,}000)$$

or

$$P = 2.0, \tag{9.6}$$

according to Equation 7.3. Then the economic profit in Equation 7.4 is

$$E = P - 1.0$$

or

$$E = 1.0 \tag{9.7}$$

Multiplying through by the cost of production, the revenue is $D = E \times C$ according to Equation 7.5, so

$$D = 1.0 \times \$20{,}000{,}000$$

or

$$D = \$20{,}000{,}000 \tag{9.8}$$

9.4.1.2 The Real Situation with No Inspection

Again, the entire production is sold at the transfer price, so

$$A = \$40{,}000{,}000 \tag{9.9}$$

The detrimental cost (value-added-detractor [VADOR]) B is the warranty cost of \$15,000 times the fraction 0.5 that fail in the 12/12 warranty period times the proportion defective p of 0.0001 times the production N of 4,000,000, so

$$B = 4{,}000{,}000 \times 0.0001 \times \$15{,}000 \times 0.5$$

or

$$B = \$3{,}000{,}000 \tag{9.10}$$

The value of C is still the production N of 4,000,000 times the production cost of \$5.00, so

$$C = 4{,}000{,}000 \times \$5.00$$

or

$$C = \$20{,}000{,}000 \tag{9.11}$$

The value of P becomes

$$P = (A - B)/C$$
$$= (\$40{,}000{,}000 - \$3{,}000{,}000)/(\$20{,}000{,}000)$$

or

$$P = 1.85 \tag{9.12}$$

making

$$E = 1.85 - 1.00$$

or

$$E = 0.85 \tag{9.13}$$

and

$$D = E \times C$$
$$= 0.85 \times \$20{,}000{,}000$$

or

$$D = \$17{,}000{,}000 \tag{9.14}$$

Without inspection, the company loses \$3 million with an error rate of only one part in 10,000.

9.4.1.3 The Real Situation with Inspection

The value of A is again the number N produced of 4,000,000 produced times the transfer price T of \$10.00, or

$$A = \$40{,}000{,}000 \tag{9.15}$$

This time some of the rods are made on overtime, but this added cost appears in the cost of production C in the denominator of P.

There is an argument between the production staff and the inspection staff as to who should absorb the cost of the nonconforming parts thrown away because of the inspection. While this author is of the opinion that the production staff should pay for the entire cost of the nonconforming parts, a contribution to B, the VADOR, is allowed to account for the proportion defective p thrown away at \$5.00 apiece. B is made

$$B = 4{,}000{,}000 \times 0.0001 \times \$5$$

or

$$B = \$2000 \tag{9.16}$$

To calculate C, one must remember that the entire batch of $N = 4{,}000{,}000$ rods is initially produced at a cost of \$5.00 apiece. Inspection must be performed, so \$200,000 is added to the cost of production. Then the rejected rods must be replaced at the overtime rate of \$10.00 each, so

$$C = 4{,}000{,}000 \times \$5.00 + 4{,}000{,}000 \times 0.0001 \times \$10.00 + \$200{,}000$$

or

$$C = \$20{,}204{,}000 \tag{9.17}$$

This makes the value of P equal to

$$P = (\$39{,}998{,}000)/(\$20{,}204{,}000)$$

or

$$P = 1.979707 \tag{9.18}$$

so that

$$E = P - 1.0$$

so

$$E = 0.979707 \tag{9.19}$$

and

$$\begin{aligned} D &= E \times C \\ &= 0.979707 \times \$20{,}204{,}000 \end{aligned}$$

or

$$D = \$19{,}794{,}000 \tag{9.20}$$

The three scenarios are summarized in Table 9.8 (Papadakis, 1996). There, one can see that a relatively small amount of nonconforming material going further into production and into the field can have a large adverse effect upon profit. The amount in this example would be \$3,000,000 lopped off the

TABLE 9.8

Comparison of Scenarios of Productivity on P/M Connecting Rods

	Scenario		
Quantity	**(a) Baseline**	**(b) No Inspection**	**(c) 100% Inspection**
P	2.00	1.85	1.98
E	1.00	0.85	0.98
D	\$20.0 M	\$17.0 M	\$19.794 M

Source: Papadakis, E.P. (1996). "Quality, Productivity, and Cash Flow." Paper 950543, Society of Automotive Engineers. Warrendale, DA, With permission.

company profit before testing. One can also see that a moderate expenditure for testing, \$200,000, can raise the profit by more than \$2,750,000 from the low of no testing. The old saw, which says that testing cannot make a profit, is not true. It would be better, of course, not to manufacture the nonconforming material, but until continuous improvement reduces the proportion of nonconforming parts, p, to a level where the calculation may show that testing is more expensive than eating the warranty cost, testing is profitable. The engine company's decision was to install and operate 100% inspection by eddy current NDT in this manufacturing situation. The engine company also insisted that the metallurgical supplier of the powder metal rod blanks carry out 100% inspection.

9.4.2 Aircraft Engine Discs

The aircraft engine discs analyzed in Section 9.3.2 using IRR are treated here by the productivity, profitability, and revenue method. The failure rates and recall campaigns postulated in Table 9.6 are used. All the data on production and costs are carried forward from Section 9.3.2. Other financial data are the transfer price T of \$5,000 and the sunk production cost at that point of \$2,000, which includes an average price of \$17.00 for the ultrasonic testing. In the calculations for comparison cases in which no testing was done, the sunk production cost would be \$1,983 on the average.

A baseline value can be run for an ideal situation of no testing and no destruction of airplanes or nacelles. In this situation, using the data in Table 9.3 where the cumulative value of N is 25,965, the value of A becomes $A = 25{,}965 \times \$5{,}000 = \$129{,}825{,}000$. One can write $B = 0$ because no parts are thrown out and no aircraft are damaged or destroyed. At the same time $C = 25{,}965 \times \$1{,}983 = \$51{,}488{,}595$. The results for P, E, and D are

$$P = (A - B)/C$$
$$= (\$129{,}825{,}000 - \$0)/(\$51{,}488{,}595)$$

so

$$P = 2.52143 \tag{9.21}$$
$$E = P - 1.0$$

so

$$E = 1.52143 \tag{9.22}$$

and

$$D = E \times C$$
$$= 1.52143 \times \$51{,}488{,}595$$

making

$$D = \$78{,}336{,}293 \tag{9.23}$$

This set of equations indicates that the engine company intended to make \$78 million dollars on the manufacturing process for jet engine discs over this production run.

Now let us consider the adverse effect of a failure of the first disc out of the posited set of failures in Table 9.6. Let us assume for the following set of calculations that an airplane is destroyed at a cost of \$7 million, and that the recall campaign follows at a cost of \$8 million. That results in B = \$15,000,000 while A is still the same at A = \$129,825,000 and C is still \$51,488,595. The results for P, E, and D are

$$P = (\$129{,}825{,}000 - \$15{,}000{,}000)/(\$51{,}488{,}595)$$

so

$$P = 2.23011 \tag{9.24}$$

$$E = P - 1.0$$

so

$$E = 1.23011 \tag{9.25}$$

and

$$D = 1.23011 \times \$51{,}488{,}595$$

making

$$D = \$63{,}336{,}636 \tag{9.26}$$

Continuing on with two planes destroyed and two recalls, three planes destroyed and three recalls, and so on, one finds the results in Table 9.9. Soon the profit expected from the manufacturing operation becomes a loss in the absence of testing. It takes only 6 out of the 8 posited nonconforming parts to cause this to happen. Recalls are not sufficient to ensure profitability.

TABLE 9.9

Productivity Calculations on Jet Engine Discs

Testing	Planes Destroyed	Campaigns	P	E	D, Dollars
No	0	0	2.52143	1.52143	78,336,405
No	1	1	2.23011	1.23011	63,336,405
No	2	2	1.93878	0.93878	48,336,405
No	3	3	1.64745	0.64745	33,336,405
No	4	4	1.35613	0.35613	18,336,405
No	5	5	1.06480	0.06480	3,336,405
No	6	6	0.77347	(0.22653)	(11,663,595)
Yes	0	0	2.49833	1.49833	77,758,422

Source: Papadakis, E. P. (1995). "Cost of Quality," *Reliability Magazine*, January/February, 8–16. With permission from Industrial Communications, Inc.

Note: Parentheses indicate negative numbers.

This onset of loss points to the need for 100% inspection. In actuality, the inspection was being done on the basis of good engineering judgment without the benefit of the financial calculations. It is instructive to perform the calculation with inspection to find the corresponding results. In the testing case, the amount shipped decreases by 34 according to Table 9.3. Thus, A becomes $25{,}931 \times \$5000 = \$129{,}655{,}000$. The value of B becomes zero because there are no accidents and no recalls. The denominator C contains the number shipped, 25,931, at their sunk cost of \$2000 including the testing cost, plus the 34 detected to have discontinuities at their sunk price of \$1000 because the testing is actually done at an intermediate stage where the cost is not fully accrued, plus the cost of testing, \$17.00, for each of the 34 discarded. C is written as $C = 25{,}931 \times \$2000 + 34 \times \$1000 + 34 \times \$17.00$, so that $C = \$51{,}896{,}578$. The values of P, E, and D are

$$P = (\$129{,}655{,}000 - \$0)/(\$51{,}896{,}578)$$

so

$$P = 2.498334 \tag{9.27}$$

$$E = P - 1.0$$

so

$$E = 1.498334 \tag{9.28}$$

and

$$\begin{aligned} D &= E \times C \\ &= 1.498334 \times \$51{,}896{,}578 \end{aligned}$$

making

$$D = \$77{,}758{,}422 \tag{9.29}$$

The resultant value of D, the dollars received from the process, is only about \$600,000 smaller than the baseline value. Hence, spending money on 100% inspection has raised the profit by a major increment vis-à-vis even the smallest number of failures of untested parts. The values in Equations 9.27 through 9.29 are entered as the last line in Table 9.9. Examination shows that inspection is profitable vs. its alternative.

9.5 Summary

The three financial calculations in this chapter provide methods for proving that 100% inspection should be performed on production. Most often the inspection methods will be NDT, which has the unique virtue of being able to detect latent defects. Depending upon the results of the financial calculations and the economic strictures of the times such as the hurdle rate determined by the chief financial officer (CFO), the results may be negative as well as positive. In other words, these same calculations can prove that the benefit–cost ratio may not be great enough to justify 100% inspection.

The first of the three methods is the DIC, which is easy to use and treats cases in which the investment is very low and the testing costs are essentially all variable costs. Several examples were given in detail showing actual examples of applications and demonstrating how to apply the DIC. The examples showed that inspection improved profitability for the companies. Repeated DIC calculations over time show that inspection continues to be needed and that continuous improvement has not caused sufficient improvement to permit the cessation of inspection.

The second method is the TARR, which is essentially equivalent to the IRR. Here a rather large investment is to be amortized over several years. Investing in inspection equipment is equivalent to investing in any other piece of equipment. It must yield a reasonable return to be justifiable. One case with a very large IRR was given in which the aircraft engine company carried on testing. Another case was shown in which the high hurdle rate at the time precluded installing a test at an automobile company even though the TARR positive and the DIC was greater than 1.0. The TARR was not great enough to exceed the company's hurdle rate at the time of the calculations.

The third method is the productivity method where the calculational trail leads from quality, to productivity, to profit, to total revenue spent to improve competitive position. It is a literal interpretation of the title of Deming's book, *Quality, Productivity, and Competitive Position* (1982). One automotive example and one aerospace example were given. Both showed that the profitability of the respective companies was decidedly enhanced by the application of 100% inspection by high-tech methods, both being NDT techniques.

In summary, the financial calculations have proved that 100% inspection raises profits and cuts costs. Inspection should continue while continuous improvement plays catch-up. It is particularly important to use NDT techniques where the nonconformities are caused by latent defects not detectable by visual inspectors, and where the nonconformities are caused by intrinsic variables measurable by high-tech correlations.

10

Nondestructive Inspection Technology and Metrology in the Context of Manufacturing Technology as Explained in This Book

10.1 Emphasis

This book has been written from my point of view as a high-tech practitioner in the realm of quality. I have noted that the basic high-tech methodologies for quality testing frequently have not been properly integrated into modern manufacturing to improve, maximize, and ensure quality. The omissions and improper utilization arise from the philosophy of manufacturing and the philosophical positions of various schools of quality management. Hence, there is a large amount of background material in the book (in chronological sequence) concerning the emphasis of different groups that have influenced quality in manufacturing.

10.2 Chronological Progression

The exposition begins with the time of cottage industry before manufacturing was even thought of, and continues through mass production. Changes after the introduction of mass production that were alleged to improve its performance are treated. Difficulties encountered are addressed. Modifications made along the way to improve quality are explained. Changes are attributed to outstanding advocates with definite positions and their own points of view. Many of the doctrines in manufacturing in general, and in quality within manufacturing in particular, are exactly that—points of view. The reader and practitioner must learn to discern the difference between proven techniques and ideas that are advocated and propagated by well-intentioned individuals and groups. While trends have been covered, the book does not mention every quality advocate, practitioner, or school of thought.

10.3 A Final Anecdote

My first experience with NDT occurred around 1938 when, as a child, I was frequently taken to High Bridge Park in the Bronx, New York. From High Bridge (the aqueduct with a foot path) one could see the New York Central main line from Grand Central Station, Manhattan, to Albany. The trains were fascinating. Although we did not know it at the time, my mother and I saw the self-propelled rail testing cars occasionally. Mother thought they were for fast mail delivery. I called them "Funny Face" because of the chevrons painted on the front.

In interviewing for a summer job in college in 1953, I saw the cars from a professional point of view at the Sperry Products Company in Danbury, Connecticut. I saw the Sperry Reflectoscope and other NDT equipment for the first time then, too.

In 2005, as I drove to the FEDEX facility to send my manuscript for this book to the editor, I was stopped at a railroad grade crossing. And what was the obstruction? A Sperry Rail Car testing the tracks. The new model was a modern truck fitted with railroad wheels, but the lettering on the side was unmistakable. It is interesting how events can come full circle in such a recognizable fashion.

Each school of thought has advocated its own position and has proposed its ideas to eliminate the observed failings of the previous school of thought. Each group succinctly spells out its unique style of improvement and denigrates the previous style. Each previous style defends itself and does not accept that it had failings. Each follow-up group is interested in promoting its own style, and does not leave a complete trail of documentation on the supposed poorer performance of the previous style. Where a statement is made in this book, there will be advocates of the opposite position. In this book, I have attempted to present the progression of ideas about quality, citing many schools of thought. Some statements may seem unsubstantiated, and indeed may be in the sense that I have studied these positions under the mentoring of experts who did not supply complete chapter-and-verse references. I have made a serious attempt to be rigorous and provide references.

High-tech methods of quality testing supported by financial calculations and statistical process control (SPC) are the specific advocacy position of this book. The high-tech applications supported by financial calculations are rigorous and referenced. Readers can see the original theory and the original data in refereed, archival journals. In particular, nondestructive testing methods are stressed. Certain types of nondestructive testing (NDT) are capable of finding intrinsic physical parameters and detecting sources of latent defects. The quality professional should become familiar with these nondestructive testing methods, which penetrate where statistics cannot go.

Using statistical process control prior to financial calculations is rigorous and provides the best approach available for using high-tech methods frugally and effectually. Previously, as shown in several chapters, inspection has been applied in ways that were not optimum. On the basis of inefficient applications, arguments have been made by some quality professionals that inspection should be eliminated. This book takes a rational and scientific approach to inspection in the context of manufacturing. Proof is offered that 100% testing by nondestructive methods can save money and improve profits rather than simply add expense. One can prove that NDT should be used in certain circumstances, and that it should not be used in other circumstances. In this, the contents of this book are different from all other advocacy presentations.

If the quality professionals and high-tech practitioners in the field of quality absorbed the information in this book in its entirety, manufacturing would be better for their efforts.

References

Advanced Systems and Designs, Inc. (1985). *Statistics and Control Chart Package.* Advanced Systems and Designs, Inc., Dearborn, MI.

ASM (1976). "Nondestructive Inspection and Quality Control." In *Metals Handbook*, 8th ed., Vol. 11. American Society for Metals, Metals Park, OH.

ASM (1985). *Metals Handbook*, 9th ed., Vol. 9. American Society for Metals, Metals Park, OH.

ASNT (1959). *Handbook of Nondestructive Testing* (2 vol.), ed. R. C. McMasters. American Society for Nondestructive Testing, Columbus, OH.

ASNT (1988). "Recommended Practice SNT-TC-1A." In *Personnel Qualification and Certification in Nondestructive Testing*. American Society for Nondestructive Testing, Columbus, OH.

ASNT (2005). *ASNT Publications Catalog*. American Society for Nondestructive Testing, Columbus, OH.

ASTM (1972). Special Technical Publication 505, *Acoustic Emission*, American Society for Testing and Materials, Philadelphia.

ASTM (2005). "Nondestructive Testing." In *Annual Book of ASTM Standards*, Vol. 3.3. American Society for Testing and Materials, Philadelphia.

Automotive Industry Action Group (1995). *Statistical Process Control — SPC*. Compiled by Chrysler, Ford, & General Motors Supplier Requirements Task Force. Automotive Industry Action Group, Southfield, MI.

BBN Software Products, Inc. (1986). *RS Series Quality Control Analysis 3.0*, Northbrook, IL.

Berlincourt, D. A., D. R. Curran, and H. Jaffe (1964). "Piezoelectric and Piezomagnetic Materials and Their Function in Transducers." In *Physical Acoustics: Principles and Method*, Vol. 1A, ed. W. P. Mason. Academic Press, New York.

Bloss, D. W. (1985). Personal communication. General Motors B.O.C., Powertrain Factory 31, Flint, MI.

Bobbin, J. E. (1974). Unpublished talk at Detroit, Michigan, chapter of ASNT. Branson Instruments, Inc., Danbury, CT.

Bray, F. (1990). Personal communication. Garrett Engine Division, Allied Signal Corp., Phoenix, AZ.

Cady, W. G. (1946). *Piezoelectricity*. McGraw-Hill, New York.

Chapman, G. B. II (1982a). "Practical NDI for Fiber-Reinforced Plastics," *Materials Engineering*, October, 72–73.

Chapman, G. B. II (1982b). "Nondestructive Inspection for Quality Assurance of Fiber-Reinforced Plastic Assemblies," Paper No. 820226. *SAE Transactions*, 91, 887–896.

Chapman, G. B. II (1983). *A Nondestructive Method of Evaluating Adhesive Bond Strength in Fiberglass Reinforced Plastic Assemblies*, STP 749. American Society for Testing and Materials, Philadelphia, Pennsylvania, pp. 32–60.

Chapman, G. B. II (1990). "Quality Systems for Automotive Plastics." In *Composite Material Technology — Processes and Properties*, eds. P. K. Mallick and S. Newman. Hanser Publishers, Vienna and New York, pp. 349–393.

Chapman, G. B. II (1991). "Methods for Testing Adhesive Bonds." In *Nondestructive Testing Handbook*, 2nd ed., Vol. 7, ed. P. McIntire. American Society for Nondestructive Testing, Columbus, OH, pp. 659–666.

Chapman, G. B. II (2004). "Infra-Red Monitoring of Friction Welds and Adhesive Bond Curing in Automotive Manufacturing." In *Proceedings of the 16th World Conference on Nondestructive Testing*, Montreal, Canada, August 30–September 3.

Chapman, G. B. II (2005a). Personal communication.

Chapman, G. B. II (2005b). "Infra-Red Monitoring of Friction Welds and Adhesive Bond Curing in Automotive Manufacturing," *CINDE Journal: Canada's National NDT Magazine* 26(3), 5–10.

Chapman, G. B. II and L. Adler (1988). *Nondestructive Inspection Technology in Quality Systems for Automotive Plastics and Composites,* Paper No. 880155. Society of Automotive Engineers, Warrendale, PA.

Chapman, G. B. II, E. P. Papadakis, and F. J. Meyer (1984). "A Nondestructive Testing Procedure for Adhesive Bonds in FRP Assemblies," *Body Engineering Journal,* Fall, 11–22.

Chippendale, T. (1996). "The Gentleman and Cabinet Maker's Director." Dover Publications, Inc. From the third London edition, 1762.

Deming, W.E. (1980). Verbal example in lecture.

Deming, W. E. (1981). Multilith notes later published as *Quality, Productivity, and Competitive Position.* MIT Center for Advanced Engineering Study, Cambridge, MA.

Deming, W. E. (1982). *Quality, Productivity, and Competitive Position.* MIT Center for Advanced Engineering Study, Cambridge, MA.

Deming, W. E. (1984). Personal communication. Ford Motor Co. Statistical Methods Council meeting, Dearborn, MI.

Eastman, H. T. (1947). Personal communication. Bradford, VT.

EH9406 (1994). "A History Lesson: The Loss of the USS *Thresher*," *Occupational Safety Observer,* June.

Enell, J. W. (1954). "What Sampling Plan Shall I Choose?," *Industrial Quality Control,* 10(6), 96–100.

Firestone, F. A. (1945a). "The Supersonic Reflectoscope for Internal Inspection," *Metals Progress,* 48, 505–512.

Firestone, F. A. (1945b). "The Supersonic Reflectoscope, an Instrument for Inspecting the Interior of Solid Parts by Means of Sound Waves," *J. Acoust. Soc. Am.*, 17, 287–300.

Firestone, F. A., and J. R. Frederick (1946). "Refinements in Supersonic Reflectoscopy: Polarized Sound," *J. Acoust. Soc. Am.*, 18, 200–211.

Ford Motor Co. (1972). "Low Carbon Bare Steel Spot Welding Schedule Standards." In *Welding Design and Reference Data WX-12*. Manufacturing Standards, Engineering and Manufacturing Staff, Ford Motor Company, Dearborn, MI, p. 3.

Ford Motor Co. (1979). *Potential Failure Mode and Effects Analysis — An Instruction Manual,* ed. Engineering and Research Staff. Ford Motor Co., Dearborn, MI.

Ford Motor Co. (1980, July). *Nondestructive Inspection (NDI) of Adhesive Bonds.* Ford Laboratory Test Method FLTM BU 17-1, Manufacturing Staff, Ford Motor Company, Dearborn, MI.

Ford Motor Co. (2005, Winter). *"My Ford" Magazine*, Time Inc. Custom Publishing, Ford Division Marketing Center, Melbourne, FL, p. 28.

Gilbreth, F. G., Jr. and E. G. Carey (1948). *Cheaper by the Dozen*. Thomas Y. Crowell, New York.

Giza, P. and E. P. Papadakis (1979). "Eddy Current Tests for Hardness Certification of Gray Iron Castings," *Materials Evaluation*, 37(8), 45–50, 55.

Granato, A. and K. Lücke (1956). "Theory of Mechanical Damping Due to Dislocations," *Journal of Applied Physics*, 27(6), 583–593.

Gray, D. E., ed. (1957). "Electricity and Magnetism," D. F. Bleil, sect. ed., In *American Institute of Physics Handbook*. McGraw-Hill, New York, Sect. 5, p. 85.

Harris, D. O. and E. Y. Lim (1983). In *Probabilistic Fracture Mechanics and Fatigue Methods: Applications for Structural Design and Maintenance*, ASTM STP 798, eds. J.M. Bloom and J.C. Ekvall. ASTM, Philadelphia, pp. 19–41.

Hildebrand, F.B. (1956). *Introduction to Numerical Analysis*, McGraw-Hill, New York, pp. 264–269.

Hoadley, B. (1986). 40th Annual ASQC Congress Transactions, Anaheim, California, May 19–21, pp. 460–466.

Hounshell, D. A. (1984). *From The American System to Mass Production, 1800–1932: The Development of Manufacturing Technology in the United States*. The Johns Hopkins University Press, Baltimore, MD.

Howell, T. (1990). Personal communication. Garrett Engine Division, Allied Signal Corp., Phoenix, AZ.

Hugo, V. (1862/1976). *Les Miserables*, trans. N. Denny. Penguin Books, New York.

IEEE (1987). ANSI/IEEE standard on piezoelectricity. Standard #176-1987. *Trans. IEEE-UFFC* , 43(5), 719–772.

Inchcape (1993). *Practical Auditing Assessment & Registration of Quality Management Systems to ISO 9000:1987; Q90:1987; BS.5750:1987*. Inchcape Testing Service, Intertek Technical Services, Fairfax, VA.

ISO (1990). Quality Systems-Model for Quality Assurance in Design, Development, Production, Installation, and Servicing. ISO/ANSI, New York.

ISO (1994a). *Quality Systems — Model for Quality Assurance in Design, Development, Production, Installation, and Servicing*, 2nd ed. International Standard ISO 9001, available through ANSI, New York.

ISO (1994b). *Quality Systems — Model for Quality Assurance in Production, Installation, and Servicing*, 2nd ed. International Standard ISO 9002, available through ANSI, New York.

ISO (2000, December 13). *Quality Management Systems — Requirements*. International Standard ANSI/ISO/ASQ Q9001-2000, American Society for Quality, Milwaukee, WI.

Jaffe, B., W. R. Cooke, and H. Jaffe (1971). *Piezoelectric Ceramics*. Academic Press, New York.

Jaffe, H. and D. A. Berlincourt (1965). "Piezoelectric Transducer Materials," *Proc. IEEE*, 53, 1372–1386.

K. J. Law Engineers, Inc. (1987). *Model 8200 Verigage Brochure*. Farmington Hills, MI.

Klenk, R. J. (1977). Personal communication. Ford Motor Co.

Kovacs, B. V. (1980). Personal communication. Ford Motor Co.

Kovacs, B. V., J. Stone, and E. P. Papadakis (1984). "Development of an Improved Sonic Resonance Inspection System for Nodularity in Crankshafts," *Materials Evaluation*, 42(7), 906–916.

Lindsay, R. B. (1960). *Mechanical Radiation*, McGraw-Hill, New York.

Lipson, C. and N. J. Sheth (1973). *Statistical Design and Analysis of Engineering Experiments*, McGraw-Hill, New York, pp. 194–224, 372–415.

Lysaght, V. E. (1949). *Indentation Hardness Testing*. Reinhold Publishing Corp., New York.

Maeva, E., I. Severina, S. Bodarenko, G. B. Chapman II, B. O'Neill, F. Sevarin, and R. G. Maev (2004). "Acoustical Methods for the Investigation of Adhesively Bonded Structures: A Review." *Canadian Journal of Physics*, 82(12), 981–1025.

Mansour, T. M. (1988). "Ultrasonic Inspection of Spot Welds in Thin Gage Steel," *Materials Evaluation*, 46(4), pp. 650–658.

Martin, B. R. (1971). *Statistics for Physicists*. Academic Press, London, pp. 85–98.

Mason, W. P. (1950). *Piezoelectric Crystals and Their Application to Ultrasonics*. Van Nostrand, New York.

Mason, W. P. (1958). *Physical Acoustics and the Properties of Solids*. Van Nostrand, New York, p. 1.

Mattiatt, O. E. (1971). *Ultrasonic Transducer Materials*. Plenum Press, New York.

McEleney, P. C. (1958). Personal communication. Watertown Arsenal Laboratories.

McMasters, R. C. (1959). *Nondestructive Testing Handbook*, Vol. 2, Sections 36–39, ed. R. C. McMasters. American Society for Nondestructive Testing, Columbus, OH.

Meeker, T. R. (1996). "Publication and proposed revision of ANSI/IEEE Standard 176-1987, ANSI/IEEE Standard on Piezoelectricity," *Trans. IEEE-UFFC*, 43(5), 717–772.

Meyer, F. J. and G. B. Chapman II (1980). "Nondestructive Testing of Bonded FRP Assemblies in the Auto Industry," *Adhesives Age*, 23(4), 21–25.

Missouri Basin Interagency Committee (1967, November). Generalized Stream Flow Probabilities — High Altitude Snow Region. Omaha, NE. Missouri Basin Interagency Committee, U.S. Army Corps of Engineers.

Mocarski, S. (1983) Personal communication. Manufacturing Development Center, Ford Motor Company, Redford, MI.

Papadakis, E. P. (1964). "Ultrasonic Attenuation and Velocity in Three Transformation Products in Steel," *J. Appl. Phys.*, 35, 1474–1482.

Papadakis, E. P. (1972). "Absolute Accuracy of the Pulse-Echo-Overlap Method and the Pulse Superposition Method for Ultrasonic Velocity," *J. Acoust. Soc. Amer.*, 52 (Pt. 2), 850–857.

Papadakis, E. P. (1974). Unpublished lecture notes from a Kepner-Tregoe lecture on the effects of changes upon processes.

Papadakis, E. P. (1975). "Ultrasonic Diffraction from Single Apertures with Application to Pulse Measurements and Crystal Physics." In *Physical Acoustics: Principles and Methods*, Vol. XI, eds. W. P. Mason and R. N. Thurston. Academic Press, New York, pp. 151–211.

Papadakis, E. P. (1976a). "Future Growth of Nondestructive Evaluation," *IEEE Trans. SU*-23(5), 284–287.

Papadakis, E. P. (1976b). "Ultrasonic Velocity and Attenuation: Measurement Methods with Scientific and Industrial Applications." In *Physical Acoustics: Principles and Methods*, Vol. XII, eds. W. P. Mason and R. N. Thurston. Academic Press, New York, pp. 277–374.

Papadakis, E. P. (1981a). "Challenges and Opportunities for Nondestructive Inspection Technology in the High-Volume Durable Goods Industry," *Materials Evaluation*, 39(2), 122–130.

Papadakis, E. P. (1981b). "Empirical Study of Acoustic Emission Statistics from Ceramic Substrates for Catalytic Converters," *Acoustica*, 48(5), 335–338.

Papadakis, E. P. (1982). "Sampling Plans and 100% Nondestructive Testing Compared," *Quality Progress*, April 38–39.

Papadakis, E. P. (1985). "The Deming Inspection Criterion for Choosing Zero or 100 Percent Inspection," *J. Quality Technology,* 17(3), 121–127.

Papadakis, E. P. (1990). "A Computer-Automated Statistical Process Control Method with Timely Response," *Engineering Costs and Production Economics,* 18, 301–310.

Papadakis, E. P. (1991). "Beam Divergence." In *Nondestructive Testing Handbook,* 2nd ed., Vol. 7, Section 3, Part 5, eds. A. S. Birks, R. E. Green, Jr., and P. McIntire. American Society for Nondestructive Testing, Columbus, OH, pp. 52–63.

Papadakis, E. P. (1992). "Inspection Decisions Based on Costs Averted," *Materials Evaluation,* 50(6), 774–776.

Papadakis, E. P. (1993). "Correlations and Functions for Determining Nondestructive Tests for Material Properties," *Materials Evaluation,* 51(5), 601–606.

Papadakis, E. P. (1995). "Cost of Quality," *Reliability Magazine,* January/February, 8–16.

Papadakis, E. P. (1996). *Quality, Productivity, and Cash Flow.* Paper No. 960543, Society of Automotive Engineers, Warrendale, PA.

Papadakis, E. P. (1997a). "Ultrasonic Instruments for Nondestructive Testing." In *Encyclopedia of Acoustics,* Vol. 2, ed. Malcolm J. Crocker. John Wiley & Sons, Inc., New York, pp. 683–693.

Papadakis, E. P. (1997b). "A Cost of Quality: Three Financial Methods for Making Inspection Decisions," *Materials Evaluation* 55(12), 1336–1345.

Papadakis, E. P. (1999). "Nondestructive Testing." In *Ultrasonic Instruments and Devices: Reference for Modern Instrumentation, Techniques, and Technology,* ed. E. P. Papadakis. Academic Press, Harcourt Science and Technology, San Diego, CA, pp. 193–274.

Papadakis, E. P. (2000a). "Troubleshooting with Failure Modes and Effects Analysis," *Materials Evaluation,* 58(4), 529–530.

Papadakis, E. P. (2000b). "Testing for Adhesive Bonding: the Problem, the Solution, and the Nationwide Fix," *Materials Evaluation,* 58(9), 1031–1034.

Papadakis, E. P. (2001). "Spot Weld Testing Dilemma: or, Good NDT, Poor Engineering," *Materials Evaluation,* 59(4), 479–480.

Papadakis, E. P. (2002). "Penny Wise, Pound Foolish: The Dangers of Skimping on NDT," *Materials Evaluation,* 60(11), 1292–1293.

Papadakis, E. P. and G. B. Chapman II (1991). "Quantitative Nondestructive Evaluation of Adhesive Lap Joints in Sheet Molding Compound by Adaptation of a Commercial Bond Tester." In *International Advances in Nondestructive Testing,* Vol. 16, ed. W. J. McGonnagle. Gordon and Breach, Philadelphia, pp. 291–330.

Papadakis, E. P. and K. A. Fowler (1972). "Observation and Analysis of Simulated Acoustic Emission Waves in Plates and Complex Structures." In *Acoustic Emission.* ASTM STP 505, American Society for Testing and Materials, Philadelphia, pp. 222–237.

Papadakis, E. P. and B. V. Kovacs (1980). "Theoretical Model for Comparison of Sonic-Resonance and Ultrasonic-Velocity Techniques for Assuring Quality in Instant Nodular Iron Parts," *Materials Evaluation,* 38(6), 25–30.

Papadakis, E. P. and R. T. Mack (1997). "Will Artificial and Human Intelligence Compete in NDT?" *Materials Evaluation,* 55(5), pp. 570–572.

Papadakis, E. P., L. Bartosiewicz, J. D. Altstetter, and G. B. Chapman II (1984). "Morphological Severity Factor for Graphite Shape in Cast Iron and Its Relation to Ultrasonic Velocity and Tensile Properties," *AFS Trans.* 92, paper #83-102, 721–728.

Papadakis, E. P., H. L. Chesney, and R. G. Hurley (1984). "Quality Assurance of Aluminum Radiators by Infrared Thermography," *Materials Evaluation*, 42(3), 333–336.

Papadakis, E. P., C. H. Stephan, M. T. McGinty, and W. B. Wall (1988). "Inspection Decision Theory: Deming Inspection Criterion and Time-Adjusted Rate-of-Return Compared," *Engineering Costs and Production Economics*, 13, 111–124.

Papadakis, E. P., C. G. Oakley, A. Selfridge, and B. Maxfield (1999). "Fabrication and Characterization of Transducers." In *Ultrasonic Instruments and Devices: Reference for Modern Instrumantation, Techniques, and Technology*, ed. E. P. Papadakis. Academic Press, Harcourt Science and Technology, San Diego, CA, pp. 472–563.

Papadakis, M. P. (2005). Personal communication.

Papadakis, P. E. (1975). Personal communication. Creighton University, Omaha, NE.

Parkhurst, F. A. (1917). *Applied Methods of Scientific Management*, 2nd ed. John Wiley & Sons, Inc., New York.

Perceptron, Inc. (1988). *Data Cam 2.0.* Farmington Hills, MI.

Plenard, E. (1964). *The Elastic Behavior of Cast Iron*. 1964 National Metal Congress, Cleveland, OH.

Rotary International (1905). Evanston, Illinois.

Scherkenbach, W. W. (1986). *The Deming Route to Quality and Productivity.* CEEP Press, Washington, D.C., pp. 60, 105.

Shewhart, W. A. (1931). *Economic Control of Quality of Manufactured Product.* Van Nostrand, New York.

Stephan, C. H. (1983). "Computer-Controlled Eddy Current Inspection of Axle Shafts for Heat Treatment." In *Computer-Integrated Manufacturing*, Vol. 8, eds. M. R. Martinez and M. C. Leu. ASME, New York.

Stephan, C. H. and H. L. Chesney (1984). "Computer-Aided Measurement of Case Depth and Surface Hardness in Automobile Axle Shafts," *Materials Evaluation*, 42(13), 1612–1618.

Taylor, Frederick W. (1911). *The Principles of Scientific Management.* Harper & Brothers, New York. Reprinted by Dover Publications, Inc., Mineola, NY, 1998.

Thomas, G. B. (1953). *Calculus and Analytic Geometry.* Addison-Wesley, Reading, MA, pp. 13–21.

Torre, R. (2005). Personal communication.

Twentieth Century Fox Films (1952). *Cheaper by the Dozen*, dir. W. Lang and prod. L. Trotti. Hollywood, CA.

Walton, M. (1986a). *The Deming Management Method.* Putnam Publishing Group, New York, p. 9.

Walton, M. (1986b). *The Deming Management Method.* Putnam Publishing Group, New York, pp. 34–36.

Western Electric Co. (1956). *Statistical Quality Control Handbook.* Western Electric Co., Newark, NJ.

Related Titles

Nondestructive Evaluation: A Tool in Design, Manufacturing, and Service
Don E. Bray
ISBN: 0849326559

Nondestructive Evaluation: Theory, Techniques, and Applications
Peter J. Shull
ISBN: 0824788729

Fundamentals of Industrial Quality Control, Third Edition
Lawrence S. Aft
ISBN: 1574441515

Index

D

E

H

I

R

S

V

W

X

Y

Z